**Zurich Lectures in Advanced Mathematics**

**Edited by**



Mathematics in Zurich has a long and distinguished tradition, in which the writing of lecture notes volumes and research monographs play a prominent part. The *Zurich Lectures in Advanced Mathematics* series aims to make some of these publications better known to a wider audience. The series has three main constituents: lecture notes on advanced topics given by internationally renowned experts, graduate text books designed for the joint graduate program in Mathematics of the ETH and the University of Zurich, as well as contributions from researchers in residence at the mathematics research institute, FIM-ETH. Moderately priced, concise and lively in style, the volumes of this series will appeal to researchers and students alike, who seek an informed introduction to important areas of current research.

Published with the support of the Huber-Kudlich-Stiftung, Zürich

Yakov B. Pesin

# Lectures on partial hyperbolicity and stable ergodicity

European Mathematical Society

Author:

Yakov B. Pesin
Department of Mathematics
The Pennsylvania State University
University Park, PA 16802
USA

2000 Mathematics Subject Classification (primary; secondary): 37-xx; 58-xx.



Bibliographic information published by Die Deutsche Bibliothek

Die Deutsche Bibliothek lists this publication in the Deutsche Nationalbibliografie; detailed bibliographic data are available in the Internet at http://dnb.ddb.de.

ISBN 3-03719-003-5



Contact address:

European Mathematical Society Publishing House
Seminar for Applied Mathematics
ETH-Zentrum FLI C1
CH-8092 Zürich
Switzerland

Phone: +41 (0)1 632 34 36
Email: info@ems-ph.org
Homepage: www.ems-ph.org

Printed on acid-free paper produced from chlorine-free pulp. TCF∞
Printed in Germany

9 8 7 6 5 4 3 2 1

*To my wife, Natasha*

# Contents

# 1. Introduction

The goal of these lectures is to present a comprehensive exposition of modern partial hyperbolicity theory. They contain the core of the theory as well as outline some recent new achievements in this rapidly developing area. The material is accessible to students and non-experts who possess some basic knowledge in dynamical systems and wish to learn some new phenomena outside classical hyperbolicity. These lectures may also be of interest to experts as they provide a unified and systematic treatment of partial hyperbolicity and stable ergodicity and are unique in that.

Partial hyperbolicity is a relatively new field, just over 30 years old, but has proven to be rich in interesting ideas, sophisticated techniques and exciting applications. It appears naturally in some models in science. To illustrate this consider the FitzHugh-Nagumo partial differential equation which is used in neurobiology to model propagation of electrical impulse through the nerve membrane:

$$u_t(x,t) = \epsilon\Delta_x u(x,t) + h(u),$$

where $u(x,t) = (u_1(x,t), u_2(x,t))$ and

$$h(u_1,u_2) = (g(u_1) - bu_2, cu_1 - du_2)$$

is the *local map*. The function $g$ introduces a cubic non-linearity

$$g(u_1) = -au_1(u_1-\theta)(u_1-1).$$

We shall discuss traveling wave solutions of the FitzHugh-Nagumo equation. These are solutions of the form

$$\varphi(\xi) = \varphi(x-ct) = (\varphi_1(x-ct), \varphi_2(x-ct)),$$

where $c > 0$ is the velocity of the wave. The function $\varphi(\xi)$ satisfies the *traveling wave equation*

$$\epsilon\varphi^{''}(\xi) + c\varphi^{'}(\xi) + h(\varphi(\xi)) = 0.$$

Setting $\varphi^{'} = v$ we obtain

$$\begin{cases} \varphi^{'} = v \\ \epsilon v^{'} = -cv - h(\varphi) \end{cases}$$

By changing the function $h(\varphi)$ outside a ball $B(0,R)$ of some large radius $R$, one can obtain that $(\varphi^{'}, v^{'})\cdot(\varphi, v) < 0$. This modification of the original system guarantees that no solutions escape to infinity which is thus a repelling fixed point. This allows us to consider the equation (and the corresponding flow) on the two-dimensional sphere.

Following principles of singular perturbation theory let us change the time to "slow time" by substituting $t = \epsilon\tau$. Denote the slow time derivative by $\dot\varphi$. We have

$$\begin{cases} \dot\varphi = \epsilon v \\ \dot v = -cv - h(\varphi). \end{cases}$$

For $\epsilon = 0$, the manifold $\mathcal{C}$, defined by $v = -\frac{1}{c}h(\varphi)$, is a manifold of equilibrium points. Consider the expanded system

$$\begin{cases} \dot\varphi = \epsilon v \\ \dot v = -cv - h(\varphi) \\ \dot\epsilon = 0 \end{cases}$$

and linearize it at $\epsilon = 0$, $v = -\frac{1}{c}h(\varphi)$. The Jacobian matrix for the linearized system has eigenvalues $\lambda = -c, -c, 0, 0, 0$. It follows that for $\varepsilon = 0$ there exist a three-dimensional center manifold $\mathcal{C}^0 = \mathcal{C}$ and a two-dimensional stable manifold to it, i.e., $\mathcal{C}$ is *normally hyperbolic* (see Section 5.1 below). By the singular perturbation theory normal hyperbolicity survives: for any sufficiently small $\epsilon$ there exist a three-dimensional center manifold $\mathcal{C}^\epsilon$ and a two-dimensional stable manifold to it. One can show that the restriction of the dynamics to $\mathcal{C}^\epsilon$ is of a Morse-Smale type (see [35]).

In a more general setting one can observe partial hyperbolicity in systems described by partial differential equations possessing inertial manifolds. It often happens that the system acts as a contraction or/and expansion in directions transversal to the inertial manifold whose rates exceed the rates of contraction and expansion along this manifold. In this case the inertial manifold is normally hyperbolic.

Partial hyperbolicity can also occur when a periodic force acts on a dissipative system $f$ possessing a "strange" attractor. The resulting system is the product $f \times \mathrm{Id}$. It acts on the phase space, that is the product of the phase space for $f$ and the circle, and possesses a partially hyperbolic "strange" attractor. A small perturbation of this map often also possesses a partially hyperbolic "strange" attractor.

The structure of these lectures is as follows. In Chapter 2 we introduce the concept of partial hyperbolicity and also describe some basic examples of partially hyperbolic diffeomorphisms. In Chapter 3 we present the Mather spectrum theory for diffeomorphisms which allows one to characterize a partially hyperbolic map in terms of the spectrum of the linear operator generated by the map in the space of all continuous vector fields. Using this characterization we establish stability of partially hyperbolic maps.

In Chapters 4, 5 and 6 we discuss various aspects of stability theory for partially hyperbolic diffeomorphisms including: 1) constructions of invariant stable and unstable foliations (see Sections 4.2–4.7); 2) some criteria for integrability of the central distribution (see Section 5.3; in general, this distribution is not integrable, see Section 6.1 but it is often integrable in a weak sense, see Section 5.6); 3) stability of the central foliation under small perturbations (see Section 5.5), and 4) the branching phenomenon for intermediate foliations (see Section 6.3). We also introduce the concept of normal hyperbolicity which originated in works of Hirsch, Pugh and Shub [25, 26] and is closely related to partial hyperbolicity.

Our approach is based on an extension and adaptation to our case of a method which originated in the work of Perron [36] (see also [3], Chapter 4; the formal description of this method is given by Theorem 4.3). This method is quite powerful and can be used in various situations. We apply it to establish structural stability of Anosov maps (see Section 4.8) and to describe some interesting phenomena associated with insufficient smoothness of intermediate foliations (see Sections 6.2 and 6.3).

In Chapter 7 we discuss a crucial absolute continuity property of invariant foliations which provides a main technical tool in studying ergodic properties of partially hyperbolic systems with respect to smooth invariant probability measures. Chapter 8 is devoted to another crucial property of invariant foliations known as the accessibility property. It is necessary and in many cases sufficient to establish topological transitivity and ergodicity of the system.

In the last two chapters we outline basic elements and recent results in Pugh-Shub stable ergodicity theory with applications to skew products over Anosov maps, to Anosov flows (in particular, geodesic flows) and to frame flows on manifolds of negative curvature. In particular, we describe the surprising "Fubini's nightmare" phenomenon associated with non-absolutely continuous "pathological" foliations arising "typically" in partial hyperbolicity theory.

The majority of results presented in these lectures come with complete proofs. However, for some results, which require sophisticated techniques, we either just outline their proofs omitting technical details (but providing necessary references) or consider the proofs of some particular cases where the main idea can still be seen. For completeness of the exposition and to broaden applications we also included some results without proofs.

**Acknowledgment.** These lecture notes are based on lectures I gave at the Eidgenössische Technische Hochschule (ETH), Zurich during the Spring semester, 2003. I would like to thank ETH for hospitality and great opportunity to work on the subject. I also would like to thank Patrick Bonvin, a graduate student at ETH, who attended these lectures and helped me with preparation of the notes. I am in debt to him for the proofreading of the entire manuscript and producing all the pictures.

While working on the final version of the manuscript I asked some experts in the field to take a look at the draft. I would like to express my deep gratitude to Misha Brin, Dima Dolgopyat, Charles Pugh, Mike Shub and Amie Wilkinson for their valuable comments, which helped me substantially improve the exposition, extend the list of references, and correct some proofs.

The final version of this manuscript was written when I was visiting the Research Institute for Mathematical Science (RIMS) at Kyoto University. I thank RIMS for hospitality and for creating an excellent atmosphere for research.

## 2. The Concept of Hyperbolicity

**2.1. Complete hyperbolicity (Anosov systems).** Partial hyperbolicity theory originated in the work of Brin and Pesin, [13]. A closely related notion of normal hyperbolicity was introduced earlier by Hirsch, Pugh and Shub [25] (see also [26]).

Partial hyperbolicity emerged in attempts to extend the notion of complete hyperbolicity introduced and studied by Anosov and Sinai [1], [2] and by Smale [48]. We begin by describing complete hyperbolicity as it will guide us through the theory of partial hyperbolicity.

Let $f$ be a diffeomorphism of class $C^q$, $q \geq 1$, of a compact smooth Riemannian manifold $M$. In what follows $M$ is always assumed to be connected, without boundary and as smooth as necessary (for simplicity the reader may assume that $M$ is of class $C^\infty$). If not specified, smoothness of diffeomorphisms, distributions (subbundles), foliations, etc. is of class $C^1$.

We say that $f$ is *(completely) hyperbolic* or an *Anosov diffeomorphism* if for each $x \in M$ there exist a decomposition of the tangent space into two subspaces

$$T_xM = E^s(x) \oplus E^u(x)$$

and constants $c > 0$ and $0 < \lambda < 1 < \mu$ such that for each $x \in M$,

1. $df E^s(x) = E^s(f(x))$ and $df E^u(x) = E^u(f(x))$;
2. $\|df^n v\| \leq c\lambda^n \|v\|$ whenever $v \in E^s(x)$ and $n \geq 0$;
3. $\|df^{-n} v\| \leq c\mu^{-n} \|v\|$ whenever $v \in E^u(x)$ and $n \geq 0$.

The subspaces $E^s(x)$ and $E^u(x)$ are called *stable* and *unstable* subspaces respectively. One can show that they depend continuously on $x$ (see Proposition 2.1 below for a more general statement), and hence, the angle between $E^s(x)$ and $E^u(x)$ is uniformly bounded away from zero.

A classical example of an Anosov diffeomorphism is a hyperbolic automorphism of the torus $\mathbb{T}^n$ given by an integer matrix $A$, with $\det A = 1$, and (complex) eigenvalues $\lambda_1, \ldots, \lambda_n$ satisfying

$$|\lambda_1| \leq \cdots \leq |\lambda_k| < 1 < |\lambda_{k+1}| \leq \cdots \leq |\lambda_n|, \quad 1 \leq k < n$$

(see Figure 1). If $e_1, \ldots, e_n$ are the corresponding (complex) eigenvectors then for each $x \in \mathbb{T}^n$,

$$E^s(x) = \operatorname{span}\{e_1, \ldots, e_k\} \cap \mathbb{R}^n, \quad E^u(x) = \operatorname{span}\{e_{k+1}, \ldots, e_n\} \cap \mathbb{R}^n.$$

Here are some highlights of the theory of Anosov maps (see [29] for more details and proofs).

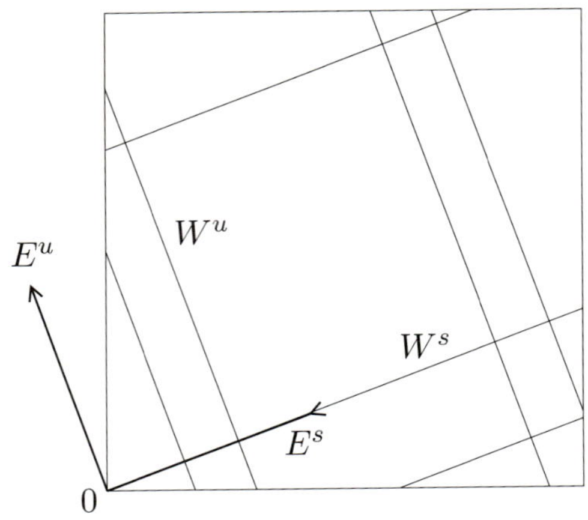


FIGURE 1. Hyperbolic automorphism of the torus $\mathbb{T}^2$.

**1. Hölder continuity of invariant distributions.** The distributions $E^s$ and $E^u$ are Hölder continuous but, in general, not smooth (this is a particular case of Theorem 3.8 below). One can show that the distribution $E^s$ (respectively, $E^u$) is smooth if it has co-dimension one.

**2. Invariant foliations.** The distributions $E^s$ and $E^u$ are integrable, their integrable manifolds form two transversal foliations $W^s$ and $W^u$ of $M$ with smooth leaves (see Figure 1; this is a particular case of Theorem 4.1 below). These foliations are invariant under $f$, i.e., for $x \in M$,

$$f(W^s(x)) = W^s(f(x)), \quad f(W^u(x)) = W^u(f(x))$$

and can be characterized as follows:

$$W^s(x) = \{y \in M\colon \rho(f^n(x), f^n(y)) \to 0,\ n \to \infty\},$$
$$W^u(x) = \{y \in M\colon \rho(f^n(x), f^n(y)) \to 0,\ n \to -\infty\}$$

where $\rho$ is the distance in $M$ induced by the Riemannian metric (one can show that convergence is, indeed, exponential). Since the distributions $E^s$ and $E^u$ are not smooth in general, the classical Frobenius integrability theorem does not apply and constructing these foliations is not a trivial problem.

**3. Topological transitivity and topological mixing.** Let $T$ be a homeomorphism of a topological space $X$. A point $x \in X$ is called *non-wandering* if for any neighborhood $U$ of $x$ there exists $n$ such that $T^n(U) \cap U \neq \emptyset$. The set of all non-wandering points is invariant and closed. The map $T$ is called *topologically transitive* if for any two open sets $U$ and $V$ there exists $n$ such that $T^n(U) \cap V \neq \emptyset$ (equivalently, there is a point $x \in X$ whose trajectory is everywhere dense). The map $T$ is called *topologically mixing* if for any two open sets $U$ and $V$ there exists $N$ such that $T^n(U) \cap V \neq \emptyset$ for each $n \geq N$.

For an Anosov diffeomorphism $f$ of a compact connected manifold the following statements are equivalent:

1. all points in $M$ are non-wandering;
2. $f$ is topologically transitive;
3. $f$ is topologically mixing;
4. each stable leaf $W^s(x)$ and each unstable leaf $W^u(x)$ is dense.

It is an open problem whether *every Anosov diffeomorphism has these properties* although in all known examples this is the case.

**4. Ergodicity.** A measure-preserving transformation $T$ of a Lebesgue space $(X, \nu)$ is called *ergodic* if any measurable invariant subset $A \subset X$ has measure either zero or one; equivalently, any measurable invariant (mod 0) function is constant (mod 0).

Assume that an Anosov diffeomorphism $f$ preserves a *smooth measure* $\nu$ on $M$ (i.e., a measure which is equivalent to the Riemannian volume). If the manifold is connected then $f$ is ergodic. If $f$ is topologically mixing then it is isometric (in the measure-theoretical sense) to a Bernoulli automorphism.

**5. Stability and structural stability.** Any sufficiently small perturbation $g$ of an Anosov diffeomorphism $f$ (in the $C^1$ topology) is again an Anosov diffeomorphism (this is a particular case of Theorem 3.6 below). Moreover, the map $g$ is topologically conjugate to $f$ via a homeomorphism $h\colon M \to M$, i.e., $f \circ h = h \circ g$ and $h$ is $C^0$ close to the identity map (see Theorem 4.11).

It follows that any sufficiently small perturbation of a linear hyperbolic automorphism of the torus is an Anosov diffeomorphism.

We describe an example of an Anosov map of algebraic origin for which the phase space is not a torus. We follow [29] (see Section 17.3).

Let $G$ be a simply connected Lie group and $\Gamma$ a (uniform) lattice, i.e., a discrete subgroup such that the factor $\Gamma \backslash G$ is compact. For some nilpotent Lie groups $\Gamma$ one can show that there exists an automorphism $F\colon G \to G$ preserving $\Gamma$, i.e., $F(\Gamma) = \Gamma$, and such that $dF|\,\mathrm{Id}$ is hyperbolic. Note that $F$ projects to $\Gamma \backslash G$. One can show that there is an invariant splitting of the Lie algebra

$$\mathfrak{L}(G) = T_{\mathrm{Id}} G = E^s \oplus E^u$$

and a norm on $\mathfrak{L}(G)$ with respect to which $dF|E^s$ and $dF^{-1}|E^u$ are both contractions. Applying the differential of the left translation $x \to gx$ one can "spread out" the invariant splitting at Id to obtain an invariant hyperbolic splitting for $F$. By construction, this splitting and the norm are invariant under left translations, so they induce a splitting and a norm on the compact nilmanifold $\Gamma\backslash G$. The factor $f\colon \Gamma\backslash G \to \Gamma\backslash G$ is an Anosov map (see also Section 6.1).

**6. Anosov flows.** We extend the notion of complete hyperbolicity to dynamical systems with continuous time.

Let $\varphi_t\colon M \to M$ be a $C^2$ flow on a compact smooth Riemannian manifold. It is generated by the vector field

$$X(t) = \frac{d}{dt}(\varphi_t(x))|_{t=0}.$$

The flow $\varphi_t$ is called an *Anosov flow* if for each $x \in M$ there exist a decomposition $T_xM = E^s(x) \oplus E^0(x) \oplus E^u(x)$ and constants $c > 0$ and $0 < \lambda < 1 < \mu$ such that for each $x \in M$,

1. $E^0(x)$ is the one-dimensional subspace generated by the vector field $X(t)$;
2. $d_x\varphi_t E^s(x) = E^s(\varphi_t(x))$ and $d_x\varphi_t E^u(x) = E^u(\varphi_t(x))$;
3. $\|d_x\varphi_t v\| \le c\lambda^t\|v\|$ whenever $v \in E^s(x)$ and $t \ge 0$;
4. $\|d_x\varphi_{-t} v\| \le c\mu^{-t}\|v\|$ whenever $v \in E^u(x)$ and $t \ge 0$.

A simple example of an Anosov flow is a *suspension flow* over an Anosov diffeomorphism. Let $f$ be an Anosov diffeomorphism of a compact Riemannian manifold $M$ and $H$ a smooth function on $M$ called the *roof* function. Consider the direct product $M \times [0, \infty)$ and define the flow $T_t$ in the following way. Starting with a point $(x, 0)$, set $T_t(x, 0) = (x, t)$ if $t < H(x)$ and $T_{H(x)}(x, 0) = (f(x), 0)$ (see Figure 2). The phase space $\mathcal{M} = M \times [0, 1]/_\sim$ for the suspension flow is diffeomorphic to the direct product $M \times [0, 1]$, where the sections $M \times 0$ and $M \times 1$ are identified by the action of $f$.

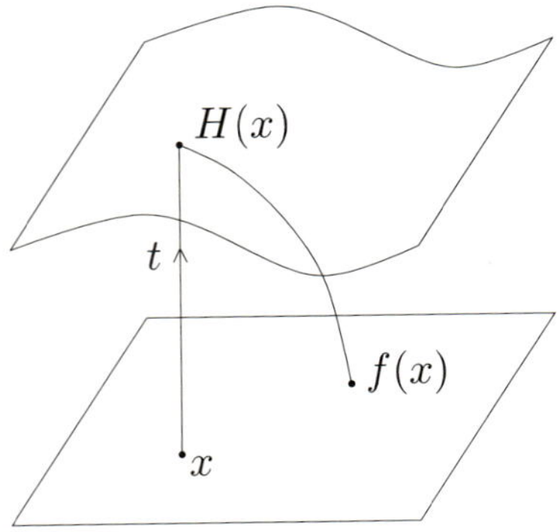


FIGURE 2. A suspension flow.

Note that if the roof function $H(x) = 1$ for all $x \in M$, the time-1 map $T_1$ is not topologically transitive, since each $t$-level of $\mathcal{M}$ gets mapped into itself. Furthermore, assume that $f$ preserves an ergodic measure $\nu$ on $M$. Then the map $T_1$ preserves the product measure on $\mathcal{M}$, but it is not ergodic. Due to ergodicity of $f$ the ergodic components of $T_1$ are precisely the $t$-level sets.

Another example of an Anosov flow is the *geodesic flow* on a compact Riemannian manifolds of negative curvature. Let $V$ be a closed oriented manifold of negative sectional curvature and $M = SV$ the unit tangent bundle of $V$. Given $x \in V$ and $v \in S_xV$ denote by $\gamma_{(x,v)}(t)$ the geodesic passing through $x$ in the direction of $v$. The geodesic flow $g_t$ is a flow on $M$ defined by $g_t(x, v) = (y, w)$ where $y = \gamma_{(x,v)}(t)$ and $w = \dot{\gamma}_{(x,v)}(t)$ (see Figure 3). The celebrated theorem of Anosov [1] claims that the flow $g_t$ is an Anosov flow.

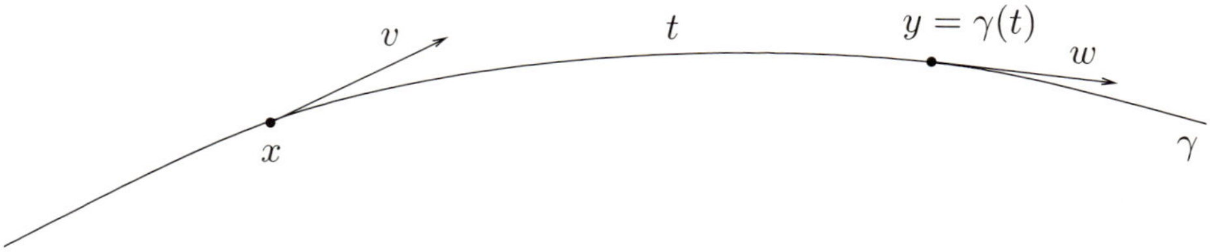


FIGURE 3. The geodesic flow.

To explain this result assume that the manifold $V$ has constant negative curvature. Consider the Poincaré (ball) model for the universal cover $\widetilde{V} = \mathbb{H}^n$ ($n$-dimensional hyperbolic space) of $V$. Given $(x, v) \in \widetilde{M} = S\widetilde{V}$, the geodesic $\gamma_{(x,v)}(t)$ determines two points, $\gamma(-\infty)$ and $\gamma(+\infty)$, on the ideal boundary of $\mathbb{H}^n$ (that is, the unit sphere). The two points $x$ and $\gamma(+\infty)$ determine the unique horosphere $H^s(x, v)$ that can be characterized as follows: for any $y \in H^s(x, v)$ there is a unique geodesic $\gamma$ passing through $y$ and asymptotic to $\gamma_{(x,v)}(t)$ (i.e., the distance $\rho(\gamma(t), \gamma_{(x,v)}(t))$ is bounded; one can show that, indeed, this distance decreases to zero with an exponential rate). The framing of $H^s(x, v)$ by the orthonormal vectors directed towards $\gamma(+\infty)$ is the global stable manifold $W^s(x, v)$ passing through $(x, v)$. Similarly, the two points $x$ and $\gamma(-\infty)$ determine the unique horosphere $H^u(x, v)$ and its framing by the orthonormal vectors directed outward $\gamma(-\infty)$ is the global unstable manifold $W^u(x, v)$ passing through $(x, v)$ (see Figure 4).

**2.2. Definition of partial hyperbolicity.** A diffeomorphism $f$ is called *partially hyperbolic* [13] if there exist numbers $0 < \lambda < \mu$, $c > 0$, and subspaces $E(x)$ and $F(x)$, $x \in M$, such that

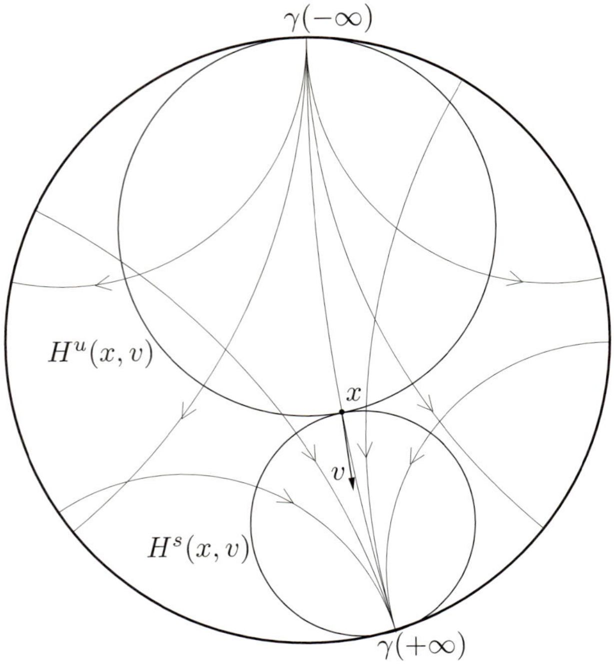


FIGURE 4. Stable and unstable horospheres for the geodesic flow.

1. $E(x)$ and $F(x)$ form an invariant splitting of the tangent space, i.e.,

$$\begin{aligned} T_xM =& E(x) \oplus F(x), \\ d_xfE(x) = E(f(x)), & \quad d_xfF(x) = F(f(x)); \end{aligned} \tag{2.1}$$

2. for $n > 0$,

$$\begin{aligned} \|d_xf^nv\| &\le c\lambda^n\|v\|, \quad v \in E(x), \\ \|d_xf^nv\| &\ge c^{-1}\mu^n\|v\|, \quad v \in F(x). \end{aligned}$$

If $\lambda < 1$ the subspace $E(x)$ is stable and it will be denoted by $E^s(x)$. If $\mu > 1$ the subspace $F(x)$ is unstable and we will use the notation $E^u(x)$.

Clearly, either $\lambda < 1$ and/or $\mu > 1$ and without loss of generality in what follows we assume the former.

The second condition characterizes the subspaces $E(x)$ and $F(x)$ in a unique way and they depend continuously in $x$; in particular, the angle between them is uniformly bounded away from zero.

**Proposition 2.1.** *The following statements hold:*

(1) *the subspace $E(x)$ consists of all vectors $v \in T_xM$ for which the estimate $\|d_xf^nv\| \le a\gamma^n\|v\|$ holds for some $a > 0$, $\lambda \le \gamma < \mu$ and all $n \ge 0$;*
(2) *the subspace $F(x)$ consists of all vectors $v \in T_xM$ for which the estimate $\|d_xf^nv\| \ge b\kappa^n\|v\|$ holds for some $b > 0$, $\lambda < \kappa \le \mu$ and all $n \ge 0$;*
(3) *the distributions $E(x)$ and $F(x)$ are continuous in $x \in M$;*
(4) *there exists $k > 0$ such that $\angle(E(x), F(x)) \ge k$ for all $x \in M$.*

*Proof.* To prove the first statement consider a vector $v \in T_xM$ and write it as $v = v_E + v_F$ with $v_E \in E(x)$ and $v_F \in F(x)$. Fix $n > 0$ and let $v_n = d_xf^nv = v_{n,E} + v_{n,F}$. Since the distributions $E$ and $F$ are $df$-invariant we have that $v_{n,E} = d_xf^nv_E$ and $v_{n,F} = d_xf^nv_F$. It follows that if $v_F \neq 0$ then for all sufficiently large $n$,

$$\|v_n\| \ge \|v_{n,F}\| = \|d_xf^nv_F\| \ge c^{-1}\mu^n\|v_F\| \ge 2a\gamma^n\|v\| > a\gamma^n\|v\|$$

and the first statement follows. The proof of the second statement is similar. [1]

To prove the third statement consider a sequence $x_m$ of points in $M$ converging to a point $x \in M$. Since the Grassmanian bundle is compact the sequence of subspaces $E(x_m)$ has a convergent subsequence $E(x_{m_k}) \to \widetilde{E}(x)$. Without loss of generality we may assume that $F(x_{m_k}) \to \widetilde{F}(x)$ and also

$$\dim E(x_{m_k}) = \dim \widetilde{E}(x), \quad \dim F(x_{m_k}) = \dim \widetilde{F}(x).$$

Fix $n > 0$. For each $k$ and $v \in E(x_{m_k})$ we have $\|df^n_{x_{m_k}}v\| \le c\lambda^n\|v\|$. Therefore, $\|df^n_{x_{m_k}}|E(x_{m_k})\| \le c\lambda^n$. Taking the limit as $k \to \infty$ yields $\|df^n_x|\widetilde{E}(x)\| \le c\lambda^n$. This implies that $\widetilde{E}(x) \subset E(x)$. Similarly, one has that $\widetilde{F}(x) \subset F(x)$ and the desired result follows from the first statement since $\dim \widetilde{E}(x) + \dim \widetilde{F}(x) = \dim M$. The last statement is an immediate corollary of the third one. □

We shall see below that the distributions $E(x)$ and $F(x)$ are, indeed, Hölder continuous.

The splitting (2.1) is unique in the following sense.

**Proposition 2.2.** *There exists $\varepsilon > 0$ with the following property. Let $\widetilde{E} \subset TM$ be an invariant distribution for which*

$$\dim \widetilde{E}(x) = \dim E(x), \quad \angle(\widetilde{E}(x), E(x)) \le \varepsilon$$

[1] I would like to thank O. Lanford who provided me with this argument.

*for every $x \in M$. Then $\widetilde{E}(x) = E(x)$. A similar statement holds for the distribution $F(x)$.*

*Proof.* Assuming the contrary, we obtain for every $v \in \widetilde{E}(x)$ that $\|d_x f^n v\| \geq a\lambda^n \|v\|$ for some $a > 0$ and all $n > 0$. By Proposition 2.1, $\widetilde{E}(x) \subset F(x)$ and hence, cannot be $\varepsilon$-close to $E(x)$. □

There is a special *Lyapunov inner metric* $<,>'$ which provides a convenient tool in the study of hyperbolicity. Choose numbers $\lambda'$ and $\mu'$ such that $0 < \lambda < \lambda' < \mu' < \mu$. Set for $v, w \in E(x)$,

$$< v, w >'_x = \sum_{k=0}^{\infty} < df^k v, df^k w >_{f^k(x)} \lambda'^{-2k}$$

and for $v, w \in F(x)$,

$$< v, w >'_x = \sum_{k=0}^{\infty} < df^{-k} v, df^{-k} w >_{f^{-k}(x)} \mu'^{2k}.$$

Notice that each series converges. Indeed, by the Cauchy-Schwarz inequality, if $v, w \in E(x)$ then

$$\begin{aligned} < v, w >'_x &\leq \sum_{k=0}^{\infty} c^2 \lambda^{2k} \lambda'^{-2k} \|v\|_x \|w\|_x \\ &= c^2 (1 - \lambda/\lambda')^{-1} \|v\|_x \|w\|_x < \infty. \end{aligned}$$

Similarly, if $v, w \in F(x)$ then

$$\begin{aligned} < v, w >'_x &\leq \sum_{k=0}^{\infty} c^2 \mu^{-2k} \mu'^{2k} \|v\|_x \|w\|_x \\ &= c^2 (1 - \mu'/\mu)^{-1} \|v\|_x \|w\|_x < \infty. \end{aligned}$$

For any vectors $v, w \in T_x M$ set

$$< v, w >'_x = < v_E, w_E >'_x + < v_F, w_F >'_x,$$

where $v = v_E + v_F$ and $w = w_E + w_F$ with $v_E, w_E \in E(x)$ and $v_F, w_F \in F(x)$.

Let $\|\cdot\|'_x$ denote the *Lyapunov norm* in $T_x M$ induced by the Lyapunov inner metric. It is easy to see that the angle

$$\angle(E(x), F(x))' = \frac{\pi}{2}$$

and that

$$\frac{1}{\sqrt{2}} \|v\|_x \leq \|v\|'_x \leq d \|v\|_x, \tag{2.2}$$

where

$$d = ck^{-1}((1-\lambda/\lambda')^{-1} + (1-\mu'/\mu)^{-1})^{\frac{1}{2}}.$$

In addition,

$$\|df|E(x)\|' \le \lambda', \quad \|df^{-1}|F(x)\|' \le (\mu')^{-1}.$$

Indeed, for $v \in E(x)$,

$$\begin{aligned}\|dfv\|'^2 &=< dfv, dfv >'= \sum_{k=0}^{\infty} < df^k dfv, df^k dfv > \lambda'^{-2k} \\ &= \sum_{k=1}^{\infty} < df^k v, df^k v > \lambda'^{-2(k-1)} = \lambda'^2 \sum_{k=1}^{\infty} < df^k v, df^k v > \lambda'^{-2k} \\ &= \lambda'^2(\|v\|'^2 - \|v\|^2) \le \lambda'^2 \|v\|'^2.\end{aligned}$$

The proof of the second inequality is similar.

It should be noticed that due to Hölder continuity of invariant distributions $E_1$ and $E_2$ (see Theorem 3.8 below), the Lyapunov inner metric is in general, only Hölder continuous but not smooth. However, in studying local stability of trajectories, we only need hyperbolic structure along individual trajectories, and we will use a Lyapunov inner metric. On the other hand, while measuring distances between points in the manifold $M$ or submanifolds of $M$ we will use the original smooth Riemannian metric. The relations (2.2) allow us to switch between the two metrics when necessary.

Using a Lyapunov inner metric one can reformulate the notion of partial hyperbolicity in the following way. A diffeomorphism $f$ is partially hyperbolic if there exist a (not necessarily smooth) Riemannian metric $\|\cdot\|$, numbers

$$0 < \lambda_1 \le \mu_1 < \lambda_2 \le \mu_2, \quad \mu_1 < 1, \tag{2.3}$$

and an invariant splitting of the tangent bundle

$$\begin{aligned} T_xM &= E_1(x) \oplus E_2(x), \\ dfE_1(x) = E_1(f(x)), &\quad dfE_2(x) = E_2(f(x)) \end{aligned} \tag{2.4}$$

such that for every $x$ the subspaces $E_1(x)$ and $E_2(x)$ are orthogonal and

$$\lambda_1 \le \|df|E_1(x)\| \le \mu_1, \quad \lambda_2 \le \|df|E_2(x)\| \le \mu_2. \tag{2.5}$$

In many situations uniform partial hyperbolicity arises in the form in which the tangent space $T_xM$ is split into three invariant distributions $E^s(x)$, $E^c(x)$ and $E^u(x)$, and the diffeomorphism contracts uniformly in the stable direction $E^s(x)$, expands uniformly in the unstable direction $E^u(x)$, and may contract or expand in the central direction $E^c(x)$ with smaller rates. More precisely, a diffeomorphism $f$ is called *partially hyperbolic in the narrow sense* if there exist numbers $C > 0$,

$$0 < \lambda_1 \le \mu_1 < \lambda_2 \le \mu_2 < \lambda_3 \le \mu_3, \quad \mu_1 < 1, \quad \lambda_3 > 1 \tag{2.6}$$

and an invariant splitting

$$\begin{aligned} T_x M &= E^s(x) \oplus E^c(x) \oplus E^u(x), \\ d_x f E^\tau(x) &= E^\tau(f(x)), \quad \tau = s, c, u \end{aligned} \tag{2.7}$$

such that for $n > 0$,

$$\begin{aligned} C^{-1}{\lambda_1}^n \|v\| \le \|d_x f^n v\| \le C{\mu_1}^n \|v\|, \quad v \in E^s(x), \\ C^{-1}{\lambda_2}^n \|v\| \le \|d_x f^n v\| \le C{\mu_2}^n \|v\|, \quad v \in E^c(x), \\ C^{-1}{\lambda_3}^n \|v\| \le \|d_x f^n v\| \le C{\mu_3}^n \|v\|, \quad v \in E^u(x). \end{aligned}$$

One can construct a Lyapunov inner metric with respect to which partial hyperbolicity in the narrow sense can be defined as follows: there exist numbers $\lambda_i, \mu_i$, $i = 1, 2, 3$ satisfying (2.3) and invariant splitting (2.4) such that for every $x \in M$, the subspaces $E^s(x)$, $E^c(x)$ and $E^u(x)$ are mutually orthogonal and

$$\begin{aligned} \lambda_1 \le \|d_x f | E^s(x)\| \le \mu_1, \quad \lambda_2 \le \|d_x f | E^c(x)\| \le \mu_2, \\ \lambda_3 \le \|d_x f | E^u(x)\| \le \mu_3. \end{aligned} \tag{2.8}$$

There is a description of partial hyperbolicity which provides a useful tool in verifying hyperbolicity conditions. Given a point $x \in M$, a subspace $E \subset T_x M$ and a number $\alpha > 0$, define the *cone* at $x$ centered around $E$ of angle $\alpha$ by

$$C(x, E, \alpha) = \{v \in T_x M \colon \angle(v, E) < \alpha\}.$$

For a partially hyperbolic diffeomorphism $f$ there are $\alpha > 0$ and two continuous cone families $C_1(x, \alpha) = C(x, E_1(x), \alpha)$ and $C_2(x, \alpha) = C(x, E_2(x), \alpha)$ for which

$$d_x f^{-1}(C_1(x, \alpha)) \subset C_1(f^{-1}(x), \alpha), \quad d_x f(C_2(x, \alpha)) \subset C_2(f(x), \alpha). \tag{2.9}$$

Moreover,

$$\begin{aligned} \|d_x f^{-1} v\| > \mu_1^{-1} \|v\|, \quad v \in C_1(x, \alpha), \\ \|d_x f v\| > \lambda_2 \|v\|, \quad v \in C_2(x, \alpha). \end{aligned} \tag{2.10}$$

It is easy to see that Conditions (2.9) and (2.10) are equivalent to partial hyperbolicity: if a diffeomorphism $f$ possesses two continuous cone families $C_1(x, \alpha) = C(x, E_1(x), \alpha)$ and $C_2(x, \alpha) = C(x, E_2(x), \alpha)$ (centered around non necessarily invariant continuous distributions $E_1(x)$ and $E_2(x)$) satisfying (2.9) and (2.10) for some $\alpha > 0$ and some $0 < \mu_1 < \lambda_2$, $\mu_1 < 1$ then $f$ is partially hyperbolic.

Similarly, partial hyperbolicity in the narrow sense can be described in terms of invariant cone families of stable and unstable cones

$$C^s(x,\alpha) = C(x, E^s(x), \alpha), \quad C^u(x,\alpha) = C(x, E^u(x), \alpha)$$

and of center-stable cones or center-unstable cones

$$C^{cs}(x,\alpha) = C(x, E^{cs}(x), \alpha), \quad C^{cu}(x,\alpha) = C(x, E^{cu}(x), \alpha).$$

Here

$$E^{cs}(x) = E^c(x) \oplus E^s(x), \quad E^{cu}(x) = E^c(x) \oplus E^u(x).$$

These cone families satisfy

$$\begin{aligned} d_x f^{-1}(C^s(x,\alpha)) &\subset C^s(f^{-1}(x),\alpha), \\ d_x f^{-1}(C^{cs}(x,\alpha)) &\subset C^{cs}(f^{-1}(x),\alpha), \\ d_x f(C^u(x,\alpha)) &\subset C^u(f(x),\alpha), \\ d_x f(C^{cu}(x,\alpha)) &\subset C^{cu}(f(x),\alpha) \end{aligned} \tag{2.11}$$

and

$$\begin{aligned} \|d_x f^{-1} v\| &> \mu_1^{-1}\|v\|, \quad v \in C^s(x,\alpha), \\ \|d_x f^{-1} v\| &> \mu_2^{-1}\|v\|, \quad v \in C^{cs}(x,\alpha), \\ \|d_x f v\| &> \lambda_3\|v\|, \quad v \in C^u(x,\alpha), \\ \|d_x f v\| &> \lambda_2\|v\|, \quad v \in C^{cu}(x,\alpha). \end{aligned} \tag{2.12}$$

Moreover, Conditions (2.11) and (2.12) are equivalent to partial hyperbolicity in the narrow sense.

We now describe the crucial Hölder continuity property of invariant distributions.

**Theorem 2.3** (Brin, [5]). *Let $f$ be a partially hyperbolic diffeomorphism in the narrow sense of a compact smooth manifold $M$. Then the distributions $E^\tau(x)$, $\tau = s, u, c, sc, uc$ are Hölder continuous , i.e., there exist $\alpha > 0$ and $C > 0$ such that*

$$\angle(E^\tau(x), E^\tau(y)) \le C\rho(x,y)^\alpha$$

*for all $x, y \in M$.*

*Proof.* We prove a more general statement below (see Theorem 3.8) □

**2.3. Examples of partially hyperbolic systems.** Here are some basic examples of partially hyperbolic diffeomorphisms.

**1. The time-$t$ map of an Anosov flow.** Let $\varphi_t$ be an Anosov flow on a compact smooth Riemannian manifold $M$. Given $t \in \mathbb{R}$, the map $\varphi_t$ is partially hyperbolic with one-dimensional central direction generated by the vector field.

**2. Frame flows.** Let $V$ be a closed oriented $n$-dimensional manifold of negative sectional curvature and $M = SV$ the unit tangent bundle of $V$. Let also $N$ be the space of positively oriented orthonormal $n$-frames in $TV$. This produces a fiber bundle $\pi\colon N \to M$ where the natural projection $\pi$ takes a frame into its first vector. The associated structure group $SO(n-1)$ acts on fibers by rotating the frames, keeping the first vector fixed. Therefore, we can identify each fiber $N_x$ with $SO(n-1)$ where $g_t$ is the geodesic flow. The *frame flow* $\Phi_t$ acts on frames by moving their first vectors according to the geodesic flow and moving the other vectors by parallel translation along the geodesic defined by the first vector. For each $t$, we have that $\pi \circ \Phi_t = g_t \circ \pi$. The frame flow $\Phi_t$ preserves the measure that is locally the product of the Liouville measure with normalized Haar measure on $SO(n-1)$. The time-$t$ map of the frame flow is a partially hyperbolic diffeomorphism (for $t \neq 0$). The center bundle has dimension $1 + \dim SO(n-1)$ and is spanned by the flow direction and the fiber direction.

**3. Direct products.** Let $f$ be an Anosov diffeomorphism of a compact manifold $M$ and $E^s_f(x)$, $E^u_f(x)$ the stable and unstable subspaces at the point $x \in M$. Let also $g$ be a diffeomorphism of a compact manifold $N$ satisfying

$$\begin{aligned}\max_{x\in M} \|df|E^s_f(x)\| &< \min_{y\in N} \|dg^{-1}(y)\|^{-1}\\ &\le \max_{y\in N} \|dg(y)\| < \min_{x\in M} \|df^{-1}|E^u_f(x)\|^{-1}.\end{aligned}$$

The map $F\colon M \times N \to M \times N$ given by $F(x,y) = (f(x), g(y))$ is partially hyperbolic. A particular case is when $g$ is the identity map of $N$ or when $N = S^1$ and $g$ is a rotation.

**4. Skew products.** Let $f$ be an Anosov diffeomorphism of a compact manifold $M$ and $g_x$, $x \in M$, a family of diffeomorphisms of a compact manifold $N$ which depend smoothly on $x$ and satisfy

$$\begin{aligned}\max_{x\in M} \|df|E^s_f(x)\| &< \min_{x\in M}\min_{y\in N} \|dg_x^{-1}(y)\|^{-1}\\ &\le \max_{x\in M}\max_{y\in N} \|dg_x(y)\| < \min_{x\in M} \|df^{-1}|E^u_f(x)\|^{-1}.\end{aligned} \tag{2.13}$$

The map $F\colon M \times N \to M \times N$ given by $F(x, y) = (f(x), g_x(y))$ is partially hyperbolic. A particular case is the map $F = F_\alpha\colon M \times S^1 \to M \times S^1$ given by

$$F(x, y) = (f(x), R_{\alpha(x)}(y)), \quad x \in M,\ y \in S^1,$$

where $R_\alpha$ is the rotation of the circle by the angle $\alpha$ and $\alpha\colon M \to M$ is a smooth function. The map $F_\alpha$ is partially hyperbolic with one-dimensional central direction.

**5. Group extensions.**An "algebraic" version of the previous example is a group extension over an Anosov diffeomorphism. Let $G$ be a compact Lie group and $\varphi\colon M \to G$ a smooth function on $M$ with values in $G$. Define the map $F = F_\varphi\colon M \times G \to M \times G$ by

$$F(x, y) = (f(x), \varphi(x)g), \quad x \in M,\ g \in G.$$

The map $F$ is partially hyperbolic since left translations are isometries of $G$ in the bi-invariant metric. If $f$ preserves a smooth probability measure $\nu$ then $F$ preserves the smooth probability measure $\nu \times \nu_G$ where $\nu_G$ is the (normalized) Haar measure on $G$.

**6. Partially hyperbolic systems on 3-dimensional manifolds.** It is an open problem to describe *compact smooth Riemannian manifolds which admit partially hyperbolic diffeomorphisms in the narrow sense.* The dimension of the manifold should be at least three. An automorphism of the torus $\mathbb{T}^3$, given by an integer matrix $A$, whose eigenvalues are $\lambda$, 1, $\lambda^{-1}$ with $|\lambda| < 1$, is partially hyperbolic. We will see later that any sufficiently small perturbation of this automorphism is partially hyperbolic in the narrow sense. In [10], Brin, Burago and Ivanov studied partiallly hyperbolic in the narrow sense diffeomorphisms on 3–dimensional manifolds.

**Theorem 2.4.** *A compact 3-dimensional manifold whose fundamental group is finite does not carry a partially hyperbolic diffeomorphism in the narrow sense whose central distribution is uniquely integrable (for the definition of unique integrability see Section 5.3).*

One can derive from this result that there are no partially hyperbolic in the narrow sense diffeomorphisms on the 3-dimensional sphere $\mathbb{S}^3$ (see [10]).

## 3. The Mather Spectrum Theory

**3.1. Mather's spectrum of a diffeomorphism.** A dynamical system $f$ on a smooth compact Riemannian manifold $M$ generates a continuous linear operator $f_*$ acting on the Banach space $\Gamma^0(TM)$ of continuous vector fields $\mathbf{v}$ on $M$ by the formula

$$(f_*\mathbf{v})(x) = df\mathbf{v}(f^{-1}(x)).$$

The spectrum $Q = Q_f$ of the complexification of $f_*$ is called *Mather's spectrum* of the dynamical system $f$. The following result provides some information on the structure of the set $Q$ as well as the corresponding structure of the tangent bundle.

**Theorem 3.1** (Mather, [31]). *Suppose that non-periodic orbits of $f$ are dense in $M$. Then*

(1) *any connected component of the spectrum $Q$ coincides with an annulus $Q_i$ around 0 with radii $\lambda_i$ and $\mu_i$ where*

$$0 < \lambda_1 \leq \mu_1 < \cdots < \lambda_t \leq \mu_t$$

*and $t \leq \dim M$;*

(2) *the invariant subspace $H_i \in \Gamma^0(TM)$ of $f_*$ corresponding to the component $Q_i$ of the spectrum is a modulus over the ring of continuous functions;*

(3) *the collection of the subspaces $E_i(x) = \{\mathbf{v}(x) \in T_x \colon \mathbf{v} \in H_i\}$ constitutes a $df$-invariant continuous distribution on $M$ and*

$$T_xM = \bigoplus_{i=1}^{t} E_i(x), \quad x \in M.$$

*Proof.* First we show that if $\mu \in Q$ and $\lambda \in \mathbb{C}$ with $|\lambda| = 1$ then $\lambda\mu \in Q$. We need the following characterization of the points in the spectrum.

**Lemma 3.2.** *$\mu \in Q$ if and only if for all $\epsilon > 0$ there exists $\mathbf{v} \in \Gamma^0(TM)$ such that $\|\mathbf{v}\| = 1$ and $\|f_*\mathbf{v} - \mu\mathbf{v}\| \leq \epsilon$.*

Let $\lambda \in \mathbb{C}$ and $|\lambda| = 1$. We wish to show that given $\epsilon > 0$, there exists $\mathbf{w} \in \Gamma^0(TM)$ such that $\|\mathbf{w}\| = 1$ and $\|f_*\mathbf{w} - \lambda\mu\mathbf{w}\| \leq \epsilon$.

Choose $\mathbf{v} \in \Gamma^0(TM)$ such that $\|\mathbf{v}\| = 1$ and $\|f_*\mathbf{v} - \mu\mathbf{v}\| \leq \epsilon/4$. Since the non-periodic points of $f$ are dense in $M$, one can choose a non-periodic point $x$ such that $\|\mathbf{v}(x)\| \geq \frac{1}{2}$. Pick a positive integer $n$ such that $n \geq \frac{4|\mu|}{\epsilon}$ and a neighborhood $U$ of $x$ such that the sets $f^k(U)$ are disjoint for $-n \leq k \leq n$.

Let $\varphi : U \to [0,1]$ be a continuous function with compact support and $\varphi(x) = 1$. Define $\psi : M \to [0,1]$ by

$$\psi(y) = \begin{cases} 0 & \text{for } y \notin \bigcup_{k=-n}^{n} f^k(U), \\ (1 - \frac{|k|}{n})\lambda^{-k}(\varphi \circ f^{-k})(y) & \text{for } y \in f^k(U). \end{cases}$$

Clearly, $\psi$ is continuous. If $\boldsymbol{\eta} = \psi \mathbf{v} \in \Gamma^0(TM)$ then

$$\|\boldsymbol{\eta}\| \geq |\boldsymbol{\eta}(x)| = |\psi(x)||\mathbf{v}(x)| \geq \frac{1}{2},$$

and

$$\begin{aligned} \|f_*\boldsymbol{\eta} - \mu\lambda\boldsymbol{\eta}\| &= \|(\psi \circ f^{-1})f_*\mathbf{v} - \mu\lambda\psi\mathbf{v}\| \\ &= \|(\psi \circ f^{-1})(f_*\mathbf{v} - \mu\mathbf{v}) + \mu(\psi \circ f^{-1})\mathbf{v} - \mu\lambda\psi\mathbf{v}\| \\ &\leq \|f_*\mathbf{v} - \mu\mathbf{v}\| + |\mu||(\psi \circ f^{-1}) - \lambda\psi||\mathbf{v}\| \\ &\leq \epsilon/4 + |\mu|/n \leq \epsilon/4 + \epsilon/4 = \epsilon/2. \end{aligned}$$

Setting $\mathbf{w} = \frac{\boldsymbol{\eta}}{\|\boldsymbol{\eta}\|}$ we conclude that $\|f_*\mathbf{w} - \lambda\mu\mathbf{w}\| \leq \epsilon$ and hence, $\lambda\mu \in Q$.

Assume now that the spectrum is split into two (not necessarily connected) components, i.e., $Q = Q_1 \cup Q_2$, $Q_1 \cap Q_2 = \emptyset$. Let $\Gamma^0(TM) = H_1 \oplus H_2$ be the corresponding splitting with $f_*(H_i) = H_i$ and $Q_i$ be the spectrum of the operator $f_*|H_i$, $i = 1, 2$.

Note that the spectral radius of the operator $f_*|H_1$ does not exceed $\mu_1$ while the spectral radius of the operator $f_*|H_2$ is at least $\lambda_2$. Therefore, $v \in H_1$ if and only if

$$\lim_n \sqrt[n]{\|f_*^n|H_1\|} \leq \mu_1$$

and $v \in H_2$ if and only if

$$\lim_n \sqrt[n]{\|f_*^n|H_2\|} \geq \lambda_2.$$

This implies that $\varphi\mathbf{v} \in H_i$ for any $\varphi \in C^0(M)$ and $\mathbf{v} \in H_i$, $i = 1, 2$, and hence, $H_1$ and $H_2$ are moduli over the ring $C^0(M)$ of continuous functions on $M$.

Finally, let

$$E_i(x) = \{\mathbf{v}(x) \in T_xM \colon \mathbf{v} \in H_i\}, \quad i = 1, 2.$$

$E_i(x)$ are clearly linear subspaces of $T_xM$ and $df E_i(x) = E_i(f(x))$. Moreover, they span $T_xM$ since any vector field $\mathbf{v}$ can be written as a sum $\mathbf{v}_1 + \mathbf{v}_2$ with $\mathbf{v}_i \in H_i$.

We shall show that $E_1(x) \cap E_2(x) = 0$. If this is not the case, then $\dim E_1(x) + \dim E_2(x) > \dim M$. Since $E_1(x)$ and $E_2(x)$ depend continuously on $x$, so do $\dim E_1(x)$ and $\dim E_2(x)$. Hence, there exists a neighborhood $U$ of $x$ such that

$$\dim E_1(y) + \dim E_2(y) = \dim E_1(x) + \dim E_2(x) > \dim M$$

for every $y \in U$. Therefore, $E_1(y) \cap E_2(y) \neq 0$ in $U$. There is a nonzero vector field $\mathbf{v} \in \Gamma^0(TU)$ with $\mathbf{v}(y) \in E_1(y) \cap E_2(y)$ for each $y \in U$. By multiplying $\mathbf{v}$ with an appropriate continuous function vanishing outside an open set $V \subset U$, we get a nonzero vector field $\mathbf{w} \in \Gamma^0(TM)$. But then $\mathbf{w} \in H_1 \cap H_2$, a contradiction. We conclude that the spectrum cannot have more than $\dim M$ connected components and the desired result follows. □

One can characterize various classes of dynamical systems using their Mather's spectra.

**Theorem 3.3** (Mather, [31]). *The following statements hold.*

(1) *A diffeomorphism $f$ is Anosov if and only if* 1 *is not contained in its Mather's spectrum $Q$.*
(2) *A diffeomorphism $f$ is partially hyperbolic if and only if its Mather's spectrum is contained in a disjoint union of two non-empty annuli, $Q \subset Q_1 \cup Q_2$ with $Q_1$ lying inside of the unit disk and/or $Q_2$ lying outside of the unit disk.*
(3) *A diffeomorphism $f$ is partially hyperbolic in the narrow sense if and only if its Mather's spectrum is contained in a disjoint union of three non-empty annuli, $Q \subset Q_1 \cup Q_2 \cup Q_3$ with $Q_1$ lying inside of the unit disk and $Q_3$ lying outside of the unit disk.*

*Proof.* We shall prove the first statement leaving the proof of other statements to the reader. It is easy to see that if $f$ is an Anosov diffeomorphism then $1 \notin Q$. It remains to show that if $1 \notin Q$ then non-periodic orbits of $f$ are dense.

Assuming the contrary there exists an open set $U \subset M$ consisting of periodic orbits. Let $P_n$ be the set of periodic orbits of period $\leq n$. These sets are closed and $U \subset \bigcup_n P_n$, Hence, one of them, say $P_m$, contains a non-empty open set $V \subset U$. Let $W \subset V$ be open and so small that its $m$ images under $f$ are disjoint. Pick a vector field $\mathbf{v} \in \Gamma^0(TM)$, $\|\mathbf{v}\| = 1$, vanishing outside $W$ and consider the vector field $\mathbf{w} = \sum_{k=0}^{m} f_*^k \mathbf{v}$. Then $f_*\mathbf{w} = \mathbf{w}$, a contradiction. □

**3.2. Stability of Mather's spectrum.** We denote by $\mathrm{Diff}^q(M)$ the space of $C^q$ diffeomorphisms equipped with the $C^q$ topology. Mather's spectrum is stable under small perturbations of dynamical systems in the following sense.

**Theorem 3.4** (Pesin, [37]). *Let $f$ be a diffeomorphism of $M$ whose non-periodic orbits are dense. Let*

$$Q_f = \bigcup_{i=1}^{t} Q_{f,i}$$

*be the decomposition of its Mather's spectrum into disjoint annuli $Q_{f,i}$ with radii*

$$0 < \lambda_{f,1} \le \mu_{f,1} < \cdots < \lambda_{f,t} \le \mu_{f,t}$$

*and $t \le \dim M$. Let also*

$$TM = \bigoplus_{i=1}^{t} E_{f,i}$$

*be the corresponding decomposition of the tangent bundle into $df$-invariant subbundles $E_{f,i}$, $i = 1, \dots, t$. Then for any sufficiently small $\varepsilon > 0$ there exists a neighborhood $\eta$ of $f$ in $\mathrm{Diff}^1(M)$ such that for any $g \in \eta$:*

(1) *Mather's spectrum $Q_g$ is a union of disjoint components $Q_{g,i}$, each being contained in an annulus with radii $\lambda_{g,i} \le \mu_{g,i}$ satisfying*

$$|\lambda_{f,i} - \lambda_{g,i}| \le \varepsilon, \quad |\mu_{f,i} - \mu_{g,i}| \le \varepsilon.$$

(2) *the distribution $E_{g,i}$ corresponding to the component $Q_{g,i}$ satisfies*

$$\max_{x \in M} \angle(E_{f,i}(x), E_{g,i}(x)) \le L\delta^\alpha \le \varepsilon,$$

*where $\delta = d_{C^1}(f, g)$ and $L > 0$, $\alpha > 0$ are constants.*

*Proof.* We begin with the following lemma.

**Lemma 3.5.** *Let $g$ be a diffeomorphism of $M$ and $\Gamma^0(TM) = H_1 \oplus H_2$ a splitting. Assume that with respect to this splitting $g_* = (G_{ji})$ where $G_{ji}\colon H_i \to H_j$, $i, j = 1, 2$ are bounded invertible linear operators satisfying*

$$\|G_{11}\| < \lambda, \quad \|G_{22}^{-1}\| < \mu, \quad \lambda\mu < 1.$$

*Then for any $\varepsilon > 0$ there is $\delta > 0$ such that if $\|G_{12}\|, \|G_{21}\| \le \delta$, then*

(1) *there is an invariant splitting $\Gamma^0(TM) = H_1' \oplus H_2'$ with respect to which $g_* = (G_{ji}')$;*
(2) *$\|(G')_{11}\| < \lambda$, $\|(G')_{22}^{-1}\| < \mu$, $(G')_{12} = 0$, $(G')_{21} = 0$;*
(3) *$d(H_i, H_i') \le L\delta^\alpha \le \varepsilon$ where $L > 0$ and $\alpha > 0$ are constants and $d$ is the distance between subspaces.*

*Proof of the lemma.* We find $H_1'$ in the form of the graph of a linear transformation $A\colon H_1 \to H_2$. Since $H_1'$ is invariant, i.e. $g_* H_1' = H_1'$, we obtain for any vector field $\mathbf{v} \in H_1$ and $x \in M$,

$$\begin{pmatrix} G_{11}(x) & G_{12}(x) \\ G_{21}(x) & G_{22}(x) \end{pmatrix} \begin{pmatrix} \mathbf{v}(x) \\ A(x)\mathbf{v}(x) \end{pmatrix} = \begin{pmatrix} G_{11}(x)\mathbf{v}(x) + G_{12}(x)A(x)\mathbf{v}(x) \\ G_{21}(x)\mathbf{v}(x) + G_{22}(x)A(x)\mathbf{v}(x) \end{pmatrix}$$
$$= \begin{pmatrix} G_{11}(x)\mathbf{v}(x) + G_{12}(x)A(x)\mathbf{v}(x) \\ A(g(x))(G_{11}(x)\mathbf{v}(x) + G_{12}(x)A(x)\mathbf{v}(x)) \end{pmatrix}.$$

Hence, the transformation $A$ should satisfy the equation

$$\begin{aligned} A(x) &= G_{22}^{-1}(x)A(g(x))G_{11}(x) \\ &\quad + G_{22}^{-1}(x)\big(-G_{21}(x) + A(g(x))G_{12}(x)A(x)\big). \end{aligned} \tag{3.1}$$

Consider the space $\mathcal{A}$ of all linear transformations $A\colon H_1 \to H_2$ for which

$$\|A\| = \sup_{x\in M} \|A(x)\| < \infty.$$

This is a Banach space with the norm $\|\cdot\|$. Consider the operator $\mathcal{Q}$ acting on $\mathcal{A}$ by the formula

$$\begin{aligned} (\mathcal{Q}A)(x) &= G_{22}^{-1}(x)A(g(x))G_{11}(x) \\ &\quad + G_{22}^{-1}(x)\big(-G_{21}(x) + A(g(x))G_{12}(x)A(x)\big). \end{aligned}$$

If $\|A\| \le 1$ and $\|A'\| \le 1$ then

$$\|\mathcal{Q}A\| \le \lambda\mu\|A\| + \mu\delta(\|A\|^2 + 1) \le \lambda\mu + 2\mu\delta = \theta$$

and

$$\|\mathcal{Q}A - \mathcal{Q}A'\| \le \lambda\mu\|A - A'\| + 2\mu\delta\|A - A'\| = \theta\|A - A'\|.$$

This shows that if $\delta$ is chosen such that $\theta < 1$ then the operator $\mathcal{Q}$ contracts the unit ball in $\mathcal{A}$ into itself. The fixed point $\widetilde{A}$ of $\mathcal{Q}$ is the unique solution of Equation (3.1). Moreover, if $\delta$ is sufficiently small then $\|\widetilde{A}\| \le \varepsilon$ and hence, $d(H_i, H_i') \le \varepsilon$.

To prove the last statement it suffices to show that $\|A\| \le L\delta^\alpha$. Fix some $0 < \delta' < \delta$ and assume that $\|G_{12}\| \le \delta'$ and $\|G_{21}\| \le \delta'$. Since the operator $\mathcal{Q}$ is contracting for any transformation $P$ with $\|P\| < 1$ we have $\|\mathcal{Q}^n P - A\| \to 0$ as $n \to \infty$. Setting $P = 0$ we get that $\mathcal{Q}0 = -G_{22}^{-1}G_{21}$ and hence, $\|\mathcal{Q}0\| \le \mu\delta'$. Proceeding by induction, we wish to show that $\|\mathcal{Q}^n 0\| \le L'\delta'$ for some constant $L' > 0$. Indeed, if $\|\mathcal{Q}^{n-1}0\| \le L'\delta'$ then

$$\begin{aligned} \|\mathcal{Q}^n 0\| &\le \|G_{22}^{-1}\mathcal{Q}^{n-1}0G_{11}\| + \|G_{22}^{-1}(-G_{21} + \mathcal{Q}^{n-1}0G_{12}\mathcal{Q}^{n-1}0)\| \\ &\le \lambda\mu L'\delta' + \mu(\delta' + (L')^2(\delta')^3). \end{aligned}$$

Thus, choosing

$$L' = \frac{\mu+1}{1-\lambda\mu}, \qquad \delta' = \frac{1}{2L'\sqrt{\mu}},$$

we obtain

$$\|\mathfrak{Q}^n 0\| \le \lambda\mu L'\delta' + \mu\delta' + \frac{1}{4}\delta' \le L'\delta'.$$

It follows that

$$\|A\| \le \lim_{n\to\infty} \left(\|\mathfrak{Q}^n 0 - A\| + \|\mathfrak{Q}^n 0\|\right) \le L'\delta'.$$

By Theorem 3.8 below $\delta' \le L''\delta^\alpha$ where $\delta = d_{C^1}(f,g)$, $L'' > 0$ is a Hölder constant and $\alpha > 0$ is a Hölder exponent of the distribution $E_{f,i}$. Therefore, $\|A\| \le L'L''\delta^\alpha$. The lemma follows. □

We proceed with the proof of the theorem. It suffices to consider the case $t = 2$ so that $\Gamma^0(TM) = H_1 \oplus H_2$, where $H_i$ are subspaces generated by $E_i$, $i = 1, 2$. Write $\tilde{g}_* = (G_{ij})$. Given $\varepsilon$, we apply Lemma 3.5 to this splitting and conclude the proof. □

The following result is an immediate corollary of Lemma 3.5.

**Theorem 3.6.** *Anosov systems, partially hyperbolic systems, and partially hyperbolic systems in the narrow sense form open subsets in* $\mathrm{Diff}^q(M)$, $q \ge 1$.

**Remark 3.7.** While a component $Q_{f,i}$ of the spectrum of $f$ may be an annulus, the corresponding component $Q_{g,i}$ of the spectrum of $g$ may consist of several annuli (see Figure 5). To illustrate this situation consider an Anosov flow $\varphi_t$ on a smooth manifold $M$ and observe that Mather's spectrum of $\varphi_0 = \mathrm{Id}$ is the unit circle while Mather's spectrum of $\varphi_t$ for $t \ne 0$ (which is partially hyperbolic in the narrow sense) contains at least two more additional annuli.

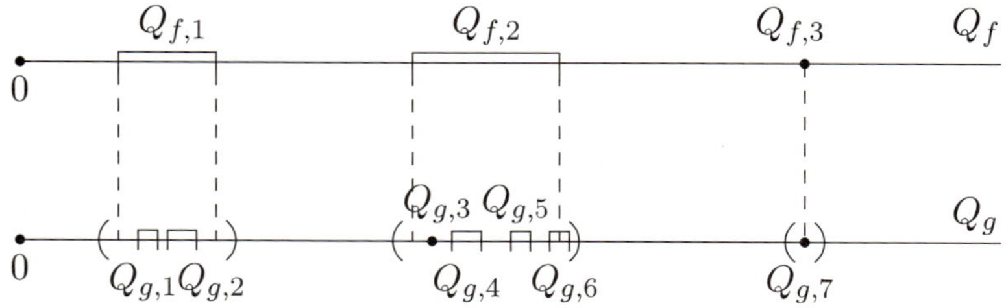


FIGURE 5. Mather's spectrum for the perturbation.

**3.3. Hölder continuity.** Consider a partially hyperbolic diffeomorphism $f$ of a compact smooth Riemannian manifold $M$ with a $df$-invariant splitting of the tangent bundle (2.4) satisfying (2.5) and (2.3).

Each distribution $E_1$ and $E_2$ is Hölder continuous. More precisely, the following statement holds.

**Theorem 3.8** (Brin, [5]). *The distribution $E_1(x)$ is Hölder continuous, i.e., for all $x, y \in M$,*

$$\angle(E_1(x), E_1(y)) \leq C\rho(x, y)^\alpha,$$

*where $C > 0$ and*

$$0 < \alpha < \frac{\log \lambda_2 - \log \mu_1}{\log \mu_2}. \tag{3.2}$$

Let us first outline the main idea of the proof without going into some technical details.

For any $\gamma > 0$ one can choose a smooth distribution $E(x)$ such that $\angle(E_1(x), E(x)) \leq \gamma$ for all $x \in M$. Since $\|df|E_1\| \leq \mu_1 < \lambda_2 \leq \|df|E_2\|$, we have that for any two distributions $E', E'' \neq E_2$,

$$\angle(f_*^{-1}E', f_*^{-1}E'') \leq \frac{\mu_1}{\lambda_2}\angle(E', E'').$$

Since $E_1$ is invariant this implies that $f_*^{-n}E \to E_1$ as $n \to \infty$. Note that the distribution $f_*^{-n}E$ is smooth for any $n \geq 0$ and in particular, is Hölder continuous. Our goal is now to show that there are positive numbers $C$ and $\alpha$ such that the Hölder continuity "survives" when passing from $f_*^{-n}E$ to $f_*^{-n-1}E$. This would imply that $E_1$ is also Hölder continuous.

Choose $\delta > 0$ so small that for any two points $x, y \in M$ with $\rho(x, y) < \delta$ there exists a unique geodesic connecting $x$ and $y$. For any distribution $E$ define the distance between subspaces $E(x)$ and $E(y)$ to be the angle $\angle(\widetilde{E}_x(y), E(y))$, where $\widetilde{E}_x(y)$ is the parallel transport of $E(x)$ from $T_xM$ to $T_yM$ along the geodesic connecting $x$ and $y$. Write

$$\begin{aligned}
\angle(f_*^{-1}E(x), f_*^{-1}E(y)) &= \angle(df^{-1}E(f(x)), df^{-1}E(f(y))) \\
&\overset{(*)}{\leq} \frac{\mu_1}{\lambda_2}\angle(E(f(x)), E(f(y))) \\
&\overset{(**)}{\leq} \frac{\mu_1}{\lambda_2}C\rho(f(x), f(y))^\alpha \\
&\leq \frac{\mu_1}{\lambda_2}C\mu_2^\alpha\rho(x, y)^\alpha.
\end{aligned}$$

Thus, the Hölder continuity of $f_*^{-1}E$ would follow if we choose $\alpha$ according to (3.2) so that

$$\frac{\mu_1}{\lambda_2}\mu_2^\alpha < 1.$$

See Figure 6. To complete the proof one should justify the inequalities $(*)$ and $(**)$. The inequality $(*)$ needs some correction since the parallel transport does not commute with the differential $df^{-1}$. The inequality $(**)$ needs some correction since the distance $\rho(f(x), f(y))$ may exceed $\delta$.

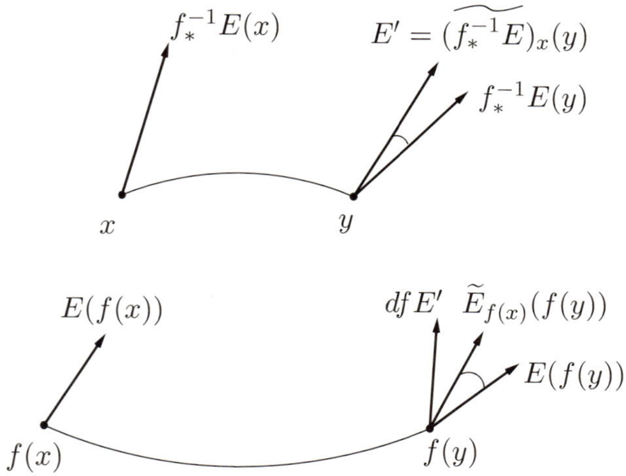


FIGURE 6. To the proof of Hölder continuity: the subspaces $dfE'$ and $\widetilde{E}_{f(x)}(f(y))$ may not coincide, and the distance $\rho(f(x), f(y))$ may exceed $\delta$.

Although these gaps can actually be fixed we prefer to present instead a more general and technically less involved proof.

By the Whitney Embedding Theorem, every manifold $M$ can be embedded in $\mathbb{R}^N$ with a sufficiently large $N$. If $M$ is compact, the Riemannian metric on $M$ is equivalent to the distance $\|x-y\|$ induced by the embedding. The Hölder exponent does not change if the Riemannian metric is changed for an equivalent smooth metric, while the Hölder constant may change. We assume, without loss of generality, that the manifold is embedded in $\mathbb{R}^N$.

For a subspace $A \subset \mathbb{R}^N$ and a vector $v \in \mathbb{R}^N$, set

$$\operatorname{dist}(v, A) = \min_{w \in A} \|v - w\|,$$

i.e., $\mathrm{dist}(v, A)$ is length of the difference between $v$ and its orthogonal projection to $A$. For subspaces $A$, $B$ in $\mathbb{R}^N$, define

$$\mathrm{dist}(A, B) = \max\left\{ \max_{v\in A, \|v\|=1} \mathrm{dist}(v, B), \max_{w\in B, \|w\|=1} \mathrm{dist}(w, A) \right\}.$$

A $k$-dimensional distribution $E$ on a set $\Lambda \subset \mathbb{R}^N$ is *Hölder continuous* with Hölder exponent $\alpha \in (0, 1]$ and Hölder constant $L > 0$ if there exists $\varepsilon_0 > 0$ such that

$$\mathrm{dist}(E(x), E(y)) \le L\|x - y\|^\alpha$$

for every $x$, $y \in \Lambda$ with $\|x - y\| \le \varepsilon_0$. The subspaces $E_1$, $E_2 \subset \mathbb{R}^N$ are said to be *$\kappa$-transverse* if $\|v_1 - v_2\| \ge \kappa$ for all unit vectors $v_1 \in E_1$ and $v_2 \in E_2$.

We obtain Theorem 3.8 from the following general result.

**Proposition 3.9** (Brin, [5]). *Let $M$ be a compact $m$-dimensional $C^2$ submanifold of $\mathbb{R}^N$ for some $m < N$, and $f\colon M \to M$ a $C^{1+\beta}$ map for some $\beta \in (0, 1)$. Assume that there exist a set $\Lambda \subset M$ and real numbers $0 < \lambda < \mu$, $c > 0$, and $\kappa > 0$ such that for each $x \in \Lambda$ there are $\kappa$-transverse subspaces $E(x)$, $F(x) \subset T_xM$ with the following properties:*

(1) *$T_xM = E(x) \oplus F(x)$;*
(2) *$\|d_xf^nv\| \le c\lambda^n\|v\|$ and $\|d_xf^nw\| \ge c^{-1}\mu^n\|w\|$ for every $v \in E(x)$, $w \in F(x)$, and every positive integer $n$.*

*Then for every $a > \max_{z\in M} \|d_zf\|^{1+\beta}$, the distribution $E$ is Hölder continuous with exponent*

$$\alpha = \frac{\log\mu - \log\lambda}{\log a - \log\lambda}\beta.$$

*Proof of the proposition.* We begin with two lemmas.

**Lemma 3.10.** *Let $A_n$ and $B_n$, for $n = 0, 1, \ldots$, be two sequences of real $N \times N$ matrices such that for some $\Delta \in (0, 1)$ and $a > 1$,*

$$\|A_n - B_n\| \le \Delta a^n$$

*for every positive integer $n$. Assume that there exist subspaces $E_A$, $E_B \subset \mathbb{R}^N$ and numbers $0 < \lambda < \mu$ and $C > 1$ such that $\lambda < a$ and for each $n \ge 0$,*

$$\|A_nv\| \le C\lambda^n\|v\| \text{ if } v \in E_A; \qquad \|A_nw\| \ge C^{-1}\mu^n\|w\| \text{ if } w \in E_A{}^\perp;$$

$$\|B_nv\| \le C\lambda^n\|v\| \text{ if } v \in E_B; \qquad \|B_nw\| \ge C^{-1}\mu^n\|w\| \text{ if } w \in E_B{}^\perp.$$

*Then*

$$\mathrm{dist}(E_A, E_B) \le 3C^2\frac{\mu}{\lambda}\Delta^{\frac{\log\mu - \log\lambda}{\log a - \log\lambda}}.$$

*Proof of the lemma.* Set

$$Q_A^n = \{v \in \mathbb{R}^N : \|A_n v\| \le 2C\lambda^n \|v\|\},$$
$$Q_B^n = \{v \in \mathbb{R}^N : \|B_n v\| \le 2C\lambda^n \|v\|\}.$$

For each $v \in \mathbb{R}^N$, write $v = v_1 + v_2$, where $v_1 \in E_A$ and $v_2 \in {E_A}^{\perp}$. If $v \in Q_A^n$, then

$$\|A_n v\| = \|A_n(v_1 + v_2)\| \ge \|A_n v_2\| - \|A_n v_1\| \ge C^{-1}\mu^n \|v_2\| - C\lambda^n \|v_1\|,$$

and hence,

$$\|v_2\| \le C\mu^{-n}(\|A_n v\| + C\lambda^n \|v_1\|) \le 3C^2 \left(\frac{\lambda}{\mu}\right)^n \|v\|.$$

Therefore,

$$\text{dist}(v, E_A) \le 3C^2 \left(\frac{\lambda}{\mu}\right)^n \|v\|. \tag{3.3}$$

Set $\gamma = \lambda/a < 1$. There exists a unique nonnegative integer $n$ such that $\gamma^{n+1} < \Delta \le \gamma^n$. If $w \in E_B$, then

$$\begin{aligned}\|A_n w\| &\le \|B_n w\| + \|A_n - B_n\| \cdot \|w\| \\ &\le C\lambda^n \|w\| + \Delta a^n \|w\| \\ &\le (C\lambda^n + (\gamma a)^n)\|w\| \le 2C\lambda^n \|w\|.\end{aligned}$$

It follows that $w \in Q_A^n$ and hence, $E_B \subset Q_A^n$. By symmetry, $E_A \subset Q_B^n$. By (3.3) and the choice of $n$, we obtain

$$\text{dist}(E_A, E_B) \le 3C^2 \left(\frac{\lambda}{\mu}\right)^n \le 3\frac{\mu}{\lambda} C^2 \Delta^{\zeta},$$

where $\zeta = \frac{\log \mu - \log \lambda}{\log a - \log \lambda}$. This completes the proof of the lemma. □

**Lemma 3.11.** *Let $f\colon M \to M$ be a $C^{1+\beta}$ map of a compact $m$-dimensional $C^2$ submanifold $M \subset \mathbb{R}^N$. Then for every $a > \max_{z\in M} \|d_z f\|^{1+\beta}$ there exists $D > 1$ such that for every $n \in \mathbb{N}$ and every $x$, $y \in M$ we have*

$$\|d_x f^n - d_y f^n\| \le D a^n \|x - y\|^{\beta}.$$

*Proof of the lemma.* Let $D'$ be such that

$$\|d_x f - d_y f\| \le D' \|x - y\|^{\beta}.$$

Set $b = \max_{z\in M} \|d_z f\| \ge 1$ and observe that for every $x$, $y \in M$,

$$\|f^n(x) - f^n(y)\| \le b^n \|x - y\|$$

Fix $a > b$. Then the lemma holds true for $n = 1$ and any $D \geq D'$. For the inductive step let us note that

$$\begin{aligned}\|d_x f^{n+1} - d_y f^{n+1}\| &\leq \|d_{f^n(x)} f\| \cdot \|d_x f^n - d_y f^n\| \\ &\quad + \|d_{f^n(x)} f - d_{f^n(y)} f\| \cdot \|d_y f^n\| \\ &\leq bDa^n \|x - y\|^\beta + D'\big(b^n \|x - y\|\big)^\beta b^n \\ &\leq Da^{n+1} \|x - y\|^\beta \left(\frac{b}{a} + \frac{D'}{D} \frac{(b^{1+\beta})^n}{a^{n+1}}\right).\end{aligned}$$

If $a > b^{1+\beta}$, then there exists $D \geq D'$ for which the factor in parentheses is less than 1. □

We proceed with the proof of the proposition.

For $x \in M$, let $(T_x M)^\perp$ denote the orthogonal complement to the tangent plane $T_x M$ in $\mathbb{R}^N$. Since $(TM)^\perp$ is a smooth distribution, it is sufficient to prove the Hölder continuity of $P = E \oplus (TM)^\perp$.

Since $E(x)$ and $F(x)$ are $\kappa$-transverse and of complementary dimensions in $T_x M$, there exists $d > 1$ such that $\|d_x f^n w\| \geq d^{-1} \mu^n \|w\|$ for every $x \in \Lambda$ and $w \perp E(x)$.

For $x$, $y \in \Lambda$ and a positive integer $n$, let $A_n$ and $B_n$ be $N \times N$ matrices such that

$$A_n v = d_x f^n v \text{ if } v \in T_x M \text{ and } A_n w = 0 \text{ if } w \in (T_x M)^\perp,$$

$$B_n v = d_y f^n v \text{ if } v \in T_y M \text{ and } B_n w = 0 \text{ if } w \in (T_y M)^\perp.$$

By Lemma 3.11,

$$\|A_n - B_n\| \leq Da^n \|x - y\|^\beta.$$

Now the proposition follows from Lemma 3.10 with $\Delta = D\|x - y\|^\beta$, $E_A = P(x)$, $E_B = P(y)$, and $C = \max\{c, d\}$. □

One should not expect the distribution $E_1$ to be smooth even in the case of Anosov diffeomorphisms. The first example of a non-smooth stable distribution was constructed by Anosov in [1]. Hasselblatt [24] has shown that for a "typical" Anosov diffeomorphism stable and unstable distributions are only Hölder continuous. Nevertheless, there are situations where these distributions are smooth:

(1) Mather's spectrum of $f$ satisfies the *pinching condition*,

$$\frac{\mu_1}{\lambda_2} \mu_2 < 1$$

(in view of (3.2) this condition guarantees that $E_1$ is Lipschitz);

(2) the distribution $E_1$ is of *co-dimension one*.

# 4. Stable and Unstable Foliations

Let $f$ be a partially hyperbolic diffeomorphism of a smooth compact Riemannian manifold $M$. Switching, if necessary, to a Lyapunov inner metric $\|\cdot\|$ (see Section 2.2) we assume that there exists a $df$-invariant decomposition of the tangent bundle

$$TM = E \oplus F \tag{4.1}$$

such that for every $x \in M$ the subspaces $E(x)$ and $F(x)$ are orthogonal and

$$\|df|E(x)\| \le \lambda, \quad \|df^{-1}|F(x)\|^{-1} \ge \mu, \tag{4.2}$$

where

$$0 < \lambda < \min\{1, \mu\}. \tag{4.3}$$

Recall that $\rho(x, y)$ denotes the distance between two points $x, y \in M$ induced by the (smooth) Riemannian metric on $M$. Given a smooth submanifold $W \subset M$, let $\rho_W(x, y)$ denotes the distance in $W$ between two points $x, y \in W$ induced by the Riemannian metric on $W$.

In this section we discuss the problem of integrability of the stable distribution $E$. Since this distribution is not in general smooth its integrability is associated with foliations of a special kind which we now consider.

**4.1. Foliations.** [13, 26].

A partition $W$ of $M$ is called a *foliation of* $M$ *with smooth leaves* or simply *foliation* if there exist $\delta > 0$ and $\ell > 0$ such that for each $x \in M$,

1. the element $W(x)$ of the partition $W$ containing $x$ is a smooth $\ell$-dimensional immersed submanifold; it is called the *(global) leaf* of the foliation at $x$; the connected component of the intersection $W(x) \cap B(x, \delta)$ that contains $x$ is called the *local leaf* at $x$ and is denoted by $V(x)$;
2. there exists a continuous map $\varphi_x\colon B(x, \delta) \to C^1(D, M)$ (where $D \subset \mathbb{R}^\ell$ is the unit ball) such that for every $y \in M \cap B(x, \delta)$ the manifold $V(y)$ is the image of the map $\varphi_x(y)\colon D \to M$.

The function $\varphi_x(y, z) = \varphi_x(y)(z)$ is called the *foliation coordinate chart.* This function is continuous and has continuous derivative $\frac{\partial}{\partial z}\varphi_x$.

A continuous distribution $E$ on $TM$ is called *integrable* if there exists a foliation $W$ of $M$ such that $E(x) = T_x W(x)$ for every $x \in M$.

**4.2. Stable Manifold theorem. The statement.** In this section we study the integrability of the stable invariant distribution $E$. If it is smooth one can use the classical Frobenius theorem to establish its integrability. Namely, one can show that if $\lambda < \mu^2$ then for any two vector fields $v, w \in E$ the Lie bracket $[v, w]$ belongs to $E$ and hence, $E$ is integrable. The corresponding foliation is smooth. Since, in general, $E$ is not smooth other methods should be exploited to establish its integrability. It can be achieved by first constructing local leaves of the foliation (local stable manifolds) and then by gluing them together.

In the hyperbolicity theory constructing local stable manifolds can be carried out by either Hadamard's method [23] (also known as the graph transfer method) or Perron's method [36]. [2] Each method is based on representing the local stable manifold as a graph of some function which satisfies an "invariance equation" (see Equation (4.12) below) and provides an approach to solve this equation. By Hadamard, the solution can be found as the fixed point of a contracting operator acting in an appropriate space of functions while by Perron, the solution can be obtained by applying the Implicit Function Theorem to a non-linear operator acting in an appropriate functional space.

Each method has its advantages and disadvantages. Hadamard's method is geometrical and is somewhat more "natural". It can also be used in non-smooth (Lipschitz) case. However, proving high smoothness of local manifolds by Hadamard's method is somewhat a difficult task. Arguably it is better to use Hadamard's method in the case of complete uniform hyperbolicity while in the case of partial hyperbolicity (as well as nonuniform hyperbolicity) Perron's method seems to be more efficient (though it is quite formal). [3]

**Theorem 4.1** (Brin and Pesin [13], Hirsch, Pugh and Shub, [26]). *(Stable Manifold Theorem). The following statements hold:*

(1) *the distribution $E$ is integrable and its maximal integral manifolds generate a foliation $W$ of $M$. The global leaf $W(x)$ which passes through the point $x \in M$ is a $C^1$-immersed submanifold of $M$;*

---

[2]Anosov remarks in [1], p. 23 (repeated by Hirsh, Pugh and Shub in [26], p. 2) "Every five years or so, if not more often, someone "discovers" the theorem of Hadamard and Perron, proving it by Hadamard's method of proof or by Perron's".

[3]Fenichel [22] used Hadamard's method and Irwin [27] used Perron's method to establish some other versions of Stable Manifold Theorem.

(2) *the foliation $W$ is $f$-invariant and contracting: for any $x \in M$, $y \in W(x)$, and $n \geq 0$,*

$$\rho_W(f^n(x), f^n(y)) \leq C(\lambda + \varepsilon)^n \rho_W(x, y), \tag{4.4}$$

*where $\varepsilon > 0$ is sufficiently small and $C = C(\varepsilon) > 0$ is a constant independent of $x$, $y$, and $n$;*

(3) *if $f$ is of class $C^q$ then so is $W(x)$ for any $x \in M$.*

**Remark 4.2.** Property (4.4) characterizes leaves of the foliation $W$, i.e., any point $y$, for which (4.4) holds for all $n \geq 0$, lies in $W(x)$. This implies that the foliation $W$ is *uniquely integrable.* In fact, it follows from Statement 3 of Theorem 4.3 that $W$ is *locally uniquely integrable.* See Section 5.3 for the definitions of unique integrability and local unique integrability.

The proof of the Stable manifold Theorem is presented in Sections 4.3–4.5. Here we briefly outline our approach. First, given a point $x \in M$, we construct a *local stable manifold* $V(x)$ (see Sections 4.3 and 4.4). This is a smooth submanifold through $x$ which consists of all points $y$ in a *small neighborhood* of $x$ whose trajectories $\{f^n(y)\}$ converge with an exponential rate to the trajectory $\{f^n(x)\}$ as $n \to +\infty$. This describes local stability properties of the trajectory $\{f^n(x)\}$. Let us stress that constructing $V(x)$ requires a partial hyperbolic structure only along the positive semi-trajectory through $x$, i.e., it does not require any other partially hyperbolic trajectories to be present. In Section 4.3 we derive an "invariance equation" for the local stable manifold, and in Section 4.4 we apply Perron's method to solve it. This can be done in an abstract setting which is described in Theorem 4.3.

In Section 4.5 we then construct the *global stable manifold* $W(x)$ at $x$ by either gluing local stable manifolds passing through different points or iterating backwards local stable manifolds along the positive semi-trajectory through $x$. The global manifold is a smooth immersed submanifold and consists of all points $y$ in the *whole* phase space $M$ whose trajectories $\{f^n(y)\}$ converge with an exponential rate to the trajectory $\{f^n(x)\}$ as $n \to +\infty$. It is the leaf of the foliation through $x$.

**4.3. The invariance equation.** We begin by constructing local stable manifolds. Given $x \in M$, a local stable manifold at $x$ can be found in the following form

$$V(x) = \exp_x\{(v, \psi_x(v)) \colon v \in B_E(r_0)\}, \tag{4.5}$$

where $B_E(r_0)$ is the ball in $E(x)$ centered at the origin of radius $r_0$ and $\psi_x \colon B_E(r_0) \to F(x)$ is a smooth map, which satisfies $\psi_x(0) = 0$ and $d\psi_x(0) =$

0. The number $r_0$ is sufficiently small and is independent of $x$. To construct the function $\psi_x$ consider the map

$$\widetilde{f}_x = \exp_{f(x)}^{-1} \circ f \circ \exp_x \colon B_E(r_0) \times B_F(r_0) \to T_{f(x)}M, \tag{4.6}$$

which is well-defined if $r_0$ is sufficiently small. Here $B_F(r_0)$ is the ball in $F(x)$ of radius $r_0$ centered at the origin. The map $\widetilde{f}_x$ can be written in the following form:

$$\widetilde{f}_x(v,w) = (A(x)v + g_x(v,w),\ B(x)w + h_x(v,w)), \tag{4.7}$$

with $v \in B_E(r_0)$ and $w \in B_F(r_0)$. Here $A(x) = df|E(x)$ and $B(x) = df|F(x)$ satisfy

$$\|A(x)\| \le \lambda, \quad (\|B(x)^{-1}\|)^{-1} \ge \mu. \tag{4.8}$$

Furthermore,

$$g_x \colon B_E(r_0) \times B_F(r_0) \to E(f(x)), \quad h_x \colon B_E(r_0) \times B_F(r_0) \to F(f(x))$$

are smooth maps satisfying

$$g_x(0,0) = 0, \quad h_x(0,0) = 0, \tag{4.9}$$

$$dg_x(0,0) = 0, \quad dh_x(0,0) = 0, \tag{4.10}$$

$$\|dg_x(v,w)\| \le C_1, \quad \|dh_x(v,w)\| \le C_1, \tag{4.11}$$

where $C_1 > 0$ is a constant.

Since the manifold $V(x)$ is invariant we have $f(V(x)) \subset V(f(x))$. Applying (4.7) yields the following equation for the function $\psi_x$:

$$B(x)\psi_x(v) + h_x(v, \psi_x(v)) = \psi_{f(x)}(A(x)v + g_x(v, \psi_x(v))). \tag{4.12}$$

Consider the trajectory $\{f^m(x)\}_{m\in\mathbb{Z}}$ and the family of maps $\widetilde{f}_{f^m(x)}$. We identify the tangent spaces $T_{f^m(x)}M$ with $\mathbb{R}^p = \mathbb{R}^\ell \times \mathbb{R}^{p-\ell}$ via an isomorphism $\tau_m$ such that $\tau_m(E(f^m(x))) = \mathbb{R}^\ell$ and $\tau_m(F(f^m(x))) = \mathbb{R}^{p-\ell}$ where $\ell = \dim E(x)$ and $p = \dim M$. Consider the map

$$F_m = \tau_{m+1} \circ \widetilde{f}_{f^m(x)} \circ {\tau_m}^{-1} \colon B^\ell(r_0) \times B^{p-\ell}(r_0) \to \mathbb{R}^p, \tag{4.13}$$

where $B^\ell(r_0)$ and $B^{p-\ell}(r_0)$ are balls in $\mathbb{R}^\ell$ and $\mathbb{R}^{p-\ell}$ respectively centered at zero of radius $r_0$. This map is of the form

$$F_m(v,w) = (A_m v + g_m(v,w), B_m w + h_m(v,w)), \tag{4.14}$$

where

$$A_m \colon \mathbb{R}^\ell \to \mathbb{R}^\ell, \quad B_m \colon \mathbb{R}^{p-\ell} \to \mathbb{R}^{p-\ell}$$

are linear maps and

$$g_m\colon B^\ell(r_0)\times B^{p-\ell}(r_0)\to\mathbb{R}^\ell,\quad h_m\colon B^\ell(r_0)\times B^{p-\ell}(r_0)\to\mathbb{R}^{p-\ell}$$

are nonlinear maps. In view of (4.8)–(4.11) these maps satisfy:

$$\|A_m\|\le\lambda,\quad (\|B_m{}^{-1}\|)^{-1}\ge\mu, \tag{4.15}$$

$$g_m(0,0)=0,\quad dg_m(0,0)=0,\quad h_m(0,0)=0,\quad dh_m(0,0)=0, \tag{4.16}$$

$$\|dg_m(v,w)\|\le C_2,\quad \|dh_m(v,w)\|\le C_2, \tag{4.17}$$

where $C_2>0$ is a constant. Using (4.14) one can rewrite Equation (4.12) in the form:

$$\psi_m(v)=B_m^{-1}\psi_{m+1}(A_mv+g_m(v,\psi_m(v)))-B_m^{-1}h_m(v,\psi_m(v)). \tag{4.18}$$

**4.4. Local stable manifold theorem.** We will now establish a general version of the Local Stable Manifold theorem which provides a unique solution of Equation (4.18).

Let us fix a number $\kappa$ such that

$$\lambda<\kappa<\mu,\quad \kappa\le 1.$$

**Theorem 4.3.** *(Local Stable Manifold Theorem). There are constants $D>0$, $0<r\le r_0$, and a map $\psi\colon B^\ell(r)\to\mathbb{R}^{p-\ell}$ such that:*

1. *$\psi$ is of class $C^1$, $\psi(0)=0$, and $d\psi(0)=0$;*
2. *if $m\ge 0$ and $v\in B^\ell(r)$ then*

$$\left(\prod_{i=0}^{m-1}F_i\right)(v,\psi(v))\in B^\ell(r_0)\times B^{p-\ell}(r_0),$$

$$\left\|\left(\prod_{i=0}^{m-1}F_i\right)(v,\psi(v))\right\|\le D\kappa^m\|(v,\psi(v))\|,$$

   *where $\prod_{i=0}^{m-1}F_i$ denotes the composition $F_{m-1}\circ\cdots\circ F_0$ with the convention that $\prod_{i=0}^{-1}F_i=\mathrm{Id}$;*
3. *given $v\in B^\ell(r)$ and $w\in B^{p-\ell}(r_0)$, if there is a number $K>0$ such that*

$$\left(\prod_{i=0}^{m-1}F_i\right)(v,w)\in B^\ell(r_0)\times B^{p-\ell}(r_0)\ \text{and}\ \left\|\left(\prod_{i=0}^{m-1}F_i\right)(v,w)\right\|\le K\kappa^m$$

   *for every $m\ge 0$, then $w=\psi(v)$;*
4. *if the map $f$ is of class $C^q$, $q\ge 1$, then so is the function $\psi$.*

**Remark 4.4.** The constants $D$ and $r$ depend only on $\lambda$, $\mu$, $\kappa$ and $C_2$ and hence, the size of local stable manifold $V(x)$ does not depend on $x \in M$.

**Remark 4.5.** The theorem is valid if instead of the finite-dimensional space $\mathbb{R}^p$ we allow a Banach space $\mathcal{B}$ with a splitting $\mathcal{B} = \mathcal{E} \oplus \mathcal{F}$ into Banach subspaces $\mathcal{E}$ and $\mathcal{F}$ and a family of local maps $F_m$ in the form (4.14) as long as the estimates (4.15) and (4.17) are uniform.

*Proof of the theorem.* Consider the linear space $\Gamma$ of sequences of vectors $z = \{z(m) \in \mathbb{R}^p\}_{m \geq 0}$ satisfying

$$\|z\|_\kappa = \sup_{m \geq 0} (\kappa^{-m} \|z(m)\|) < \infty. \tag{4.19}$$

$\Gamma$ is a Banach space with the norm $\|z\|_\kappa$. Set

$$W = \{z \in \Gamma \colon z(m) \in B^\ell(r_0) \times B^{p-\ell}(r_0) \text{ for every } m \geq 0\}. \tag{4.20}$$

Since $\kappa \leq 1$ the set $W$ is open. Consider the map $\Phi \colon B^\ell(r_0) \times W \to \Gamma$ given by

$$\Phi(y, z)(m) = -z(m) + (\Phi_1(y, z)(m), \Phi_2(y, z)(m)), \tag{4.21}$$

where

$$\Phi_1(y, z)(0) = y, \quad \Phi_2(y, z)(0) = -\sum_{n=0}^{\infty} \left( \prod_{i=0}^{n} B_i^{-1} \right) h_n(z(n)) \tag{4.22}$$

and if $m > 0$ then

$$\begin{aligned} \Phi_1(y, z)(m) &= \left( \prod_{i=0}^{m-1} A_i \right) y + \sum_{n=0}^{m-1} \left( \prod_{i=n+1}^{m-1} A_i \right) g_n(z(n)), \\ \Phi_2(y, z)(m) &= -\sum_{n=0}^{\infty} \left( \prod_{i=0}^{n} B_{i+m}^{-1} \right) h_{n+m}(z(n+m)). \end{aligned} \tag{4.23}$$

We will show that the map $\Phi$ is well-defined and is continuously differentiable in $y$ and $z$. Indeed, by (4.16) and (4.17) for $z \in B^\ell(r_0) \times B^{p-\ell}(r_0)$ and $n \geq 0$,

$$\|g_n(z)\| = \|g_n(z) - g_n(0)\| \leq \|dg_n(\xi)\| \|z\| \leq C_2 \|z\|, \tag{4.24}$$

where $\xi$ lies on the interval that joins the points 0 and $z$. Similarly, we have that

$$\|h_n(z)\| \leq C_2 \|z\|. \tag{4.25}$$

Using (4.15), (4.24) and (4.25), we find that

$$\|\Phi(y,z)\|_\kappa = \sup_{m\geq 0}(\kappa^{-m}\|\Phi(y,z)(m)\|) \leq \sup_{m\geq 0}(\kappa^{-m}\|z(m)\|)$$
$$+ \sup_{m\geq 0}\left\{\kappa^{-m}\left[\left(\prod_{i=0}^{m-1}\|A_i\|\right)\|y\| + \sum_{n=0}^{m-1}\left(\prod_{i=n+1}^{m-1}\|A_i\|\right)C_2\|z(n)\|\right.\right.$$
$$\left.\left. + \sum_{n=0}^{\infty}\left(\prod_{i=0}^{n}\|B_{i+m}^{-1}\|\right)C_2\|z(n+m)\|\right]\right\}$$
$$\leq \|z\|_\kappa + \sup_{m\geq 0}(\kappa^{-m}\lambda^m)\|y\|$$
$$+ \sup_{m\geq 0}\left(\kappa^{-m}C_2\|z\|_\kappa\left[\sum_{n=0}^{m-1}\lambda^{m-n-1}\kappa^n + \sum_{n=0}^{\infty}\mu^{-(n+1)}\kappa^{(m+n)}\right]\right).$$

Note that

$$\sup_{m\geq 0}(\kappa^{-m}\lambda^m) = 1.$$

It is easy to check that

$$\sup_{m\geq 0}\left[\kappa^{-m}\lambda^{m-1}\sum_{n=0}^{m-1}\left(\lambda^{-1}\kappa\right)^n\right] = C_3 < \infty$$

and

$$\sup_{m\geq 0}\left(\mu^{-1}\sum_{n=0}^{\infty}\left(\mu^{-1}\kappa\right)^n\right) = C_4 < \infty.$$

Setting $M = C_3 + C_4$ we conclude that

$$\|\Phi(y,z)\|_\kappa \leq \|y\| + (C_2M+1)\|z\|_\kappa.$$

This implies that the map $\Phi$ is well-defined. Clearly, $\Phi(0,0) = (0,0)$.

We will show that the map $\Phi$ is of class $C^1$. Indeed, for any $y \in B^\ell(r_0)$ and $t \in \mathbb{R}^\ell$ such that $y + t \in B^\ell(r_0)$, any $z \in W$, and $m \geq 0$ we have

$$\Phi(y+t,z)(m) - \Phi(y,z)(m) = \left(\left(\prod_{i=0}^{m-1}A_i\right)t,\ 0\right).$$

It follows that

$$\partial_y\Phi(y,z)(m) = \left(\prod_{i=0}^{m-1}A_i,\ 0\right). \tag{4.26}$$

For any $y \in B^\ell(r_0)$, $z \in W$, $t \in \Gamma$ such that $z + t \in W$ we write

$$\Phi(y,z+t) - \Phi(y,z) = (\mathcal{A}(z) - \mathrm{Id})t + o(z,t), \tag{4.27}$$

where Id is the identity map,

$$
(4.28)\quad \begin{aligned} \mathcal{A}(z)t(m) = \Bigg( & \sum_{n=0}^{m-1} \left( \prod_{i=n+1}^{m-1} A_i \right) dg_n(z(n))t(n) \, , \\ & - \sum_{n=0}^{\infty} \left( \prod_{i=0}^{n} B_{i+m}^{-1} \right) dh_{n+m}(z(n+m))t(m+n) \Bigg) \, , \end{aligned}
$$

and

$$
\begin{aligned} o(z,t)(m) = \Bigg( & \sum_{n=0}^{m-1} \left( \prod_{i=n+1}^{m-1} A_i \right) o_1(z,t)(n), \\ & - \sum_{n=0}^{\infty} \left( \prod_{i=0}^{n} B_{i+m}^{-1} \right) o_2(z,t)(m+n) \Bigg) . \end{aligned}
$$

Here $o_i(z,t)(m)$, for $i = 1, 2$, are given by

$$
\begin{aligned} o_1(z,t)(m) &= g_m((z+t)(m)) - g_m(z(m)) - dg_m(z(m))t(m), \\ o_2(z,t)(m) &= h_m((z+t)(m)) - h_m(z(m)) - dh_m(z(m))t(m). \end{aligned}
$$

If $z_1$, $z_2 \in W$ and $t \in \Gamma$ then by (4.15) and (4.17),

$$
\begin{aligned} \|\mathcal{A}(z)t\|_\kappa \le \sup_{m\ge 0} \Bigg\{ \kappa^{-m} \Bigg[ & \sum_{n=0}^{m-1} \left( \prod_{i=n+1}^{m-1} \|A_i\| \right) C_2 \|t(n)\| \\ & + \sum_{n=0}^{\infty} \left( \prod_{i=0}^{n} \|B_{i+m}^{-1}\| \right) C_2 \|t(n+m)\| \Bigg] \Bigg\} \\ \le \sup_{m\ge 0} \Bigg\{ \kappa^{-m} C_2 \Bigg[ & \sum_{n=0}^{m-1} \lambda^{m-n-1} \kappa^n + \sum_{n=0}^{\infty} \mu^{-(n+1)} \kappa^{(n+m)} \Bigg] \Bigg\} \|t\|_\kappa . \end{aligned}
$$

It follows that

$$
\|\mathcal{A}(z)t\|_\kappa \le C_2 M \|t\|_\kappa .
$$

Therefore, by the Mean Value Theorem, $\|o_i(z,t)(m)\| = o(\|t(m)\|)$. Arguing as above we find that

$$
\|o(z,t)\|_\kappa = o(\|t\|_\kappa).
$$

This implies that $\partial_z \Phi(y,z) = \mathcal{A}(z) - \mathrm{Id}$. In particular, $\partial_z \Phi(y,0) = -\mathrm{Id}$. Clearly, the map $\partial_z \Phi(y,z)$ is continuous. Therefore, the map $\Phi$ satisfies the conditions of the Implicit Function Theorem, and hence, there exist a number $0 < r \le r_0$ and a map $\varphi \colon B^\ell(r) \to W$ of class $C^1$ such that

$$
(4.29)\qquad \varphi(0) = 0 \quad \text{and} \quad \Phi(y, \varphi(y)) = 0.
$$

We describe some properties of the map $\varphi$. Differentiating the second equality in (4.29) with respect to $y$ yields

$$d\varphi(y) = -[\partial_z\Phi(y,\varphi(y))]^{-1}\partial_y\Phi(y,\varphi(y)).$$

Setting $y = 0$ in this equality yields

$$d\varphi(0)(m) = \left(\prod_{i=0}^{m-1} A_i,\ 0\right), \quad y \in B^\ell(r). \tag{4.30}$$

One can write the vector $\varphi(y)(m)$ in the form

$$\varphi(y)(m) = (\varphi_1(y)(m),\ \varphi_2(y)(m)),$$

where $\varphi_1(y)(m) \in \mathbb{R}^\ell$ and $\varphi_2(y)(m) \in \mathbb{R}^{p-\ell}$. It follows from (4.29) that if $m \geq 0$ then

$$\varphi_1(y)(m) = \left(\prod_{i=0}^{m-1} A_i\right) y + \sum_{n=0}^{m-1}\left(\prod_{i=n+1}^{m-1} A_i\right) g_n(\varphi(y)(n)) \tag{4.31}$$

and

$$\varphi_2(y)(m) = -\sum_{n=0}^{\infty}\left(\prod_{i=0}^{n} B_{i+m}^{-1}\right) h_{n+m}(\varphi(y)(n+m)). \tag{4.32}$$

These equalities imply that

$$\varphi_1(y)(m+1) = A_m\varphi_1(y)(m) + g_m(\varphi_1(y)(m), \varphi_2(y)(m)),$$
$$\varphi_2(y)(m+1) = B_m\varphi_2(y)(m) + h_m(\varphi_1(y)(m), \varphi_2(y)(m)).$$

Indeed, iterating the first equality forward one easily obtains (4.31). Rewriting the second equality in the form

$$\varphi_2(y)(m) = B_m^{-1}\varphi_2(y)(m+1) - B_m^{-1}h_m(\varphi_1(y)(m), \varphi_2(y)(m))$$

and iterating it backward yields (4.32). Therefore, the function $\varphi(y)$ is invariant under the family of maps $F_m$,

$$F_m(\varphi(y)(m)) = \varphi(y)(m+1). \tag{4.33}$$

The desired map $\psi$ is now defined by

$$\psi(v) = \varphi_2(v)(0) \tag{4.34}$$

for each $v \in B^\ell(r)$. Note that $\varphi_1(v)(0) = v$. It follows from (4.29) and (4.30) that the map $\psi$ satisfies Statement 1 of the theorem. Furthermore, by (4.33),

$$\prod_{i=0}^{m-1} F_i(v,\psi(v)) = \prod_{i=0}^{m-1} F_i(\varphi_1(v)(0),\ \varphi_2(v)(0)) = \prod_{i=0}^{m-1} F_i(\varphi(v)(0)) = \varphi(v)(m) \tag{4.35}$$

and

$$\left\|\prod_{i=0}^{m-1} F_i(v,\psi(v))\right\| \le \kappa^m\|\varphi(v)\|_\kappa = \kappa^m\|\varphi(v)-\varphi(0)\|_\kappa \le \kappa^m \sup_{\xi\in B^\ell(r)} \|d\varphi(\xi)\|_\kappa\|v\| \le D\kappa^m\|v\| \le D\kappa^m\|(v,\psi(v))\| \tag{4.36}$$

for some $D > 0$ (we use here the fact that $\|v_1\|, \|v_2\| \le \|v\|$ for every $v = (v_1, v_2) \in \mathbb{R}^p$). This proves Statement 2. Let $(v,w) \in B^\ell(r) \times B^{p-\ell}(r_0)$ satisfy the assumptions of Statement 3 of the lemma. Set

$$z(m) = \prod_{i=0}^{m-1} F_i(v,w).$$

It follows that $z \in \Gamma$ (since $\|z\|_\kappa \le K$) and that $\Phi(v,z) = 0$. The uniqueness of the map $\varphi$ implies that $z = \varphi(v)$ and hence,

$$w = \varphi_2(v)(0) = \psi(v).$$

This establishes Statement 3. To prove Statement 4 it suffices to show that the map $\Phi$ is of class $C^q$. Indeed, a simple modification of the above argument shows that

$$d_y^k\Phi(y,z) = (0,0), \quad 2 \le k \le q$$

and

$$d_z^k\Phi(y,z) = \left(\sum_{n=0}^{m-1}\left(\prod_{i=n+1}^{m-1} A_i\right) d^k g_n(z(n)), -\sum_{n=0}^{\infty}\left(\prod_{i=0}^{n} B_{i+m}^{-1}\right) d^k h_{m+n}(z(m+n))\right). \tag{4.37}$$

We leave the details to the reader. □

**4.5. Construction of global manifolds.** In this section we construct the global stable manifold at a point $x$. There are two ways to do this. The first one is by *gluing* the local stable manifolds. Namely, if $y \in V(x)$ then the local stable manifold $V(y)$ has the same "size" as $V(x)$ and is a continuation of $V(x)$, i.e., $V(x) \cup V(y)$ is a smooth submanifold in $M$. Continuing this way one will obtain the global manifold $W(x)$. We shall not discuss details of this process and proceed with another way of constructing the global stable manifolds.

It is based on *iterating* the local stable manifolds backward along the positive semi-trajectory of $x$. Namely, let

$$W(x) = \bigcup_{n=0}^{\infty} f^{-n}(V(f^n(x))). \tag{4.38}$$

It is a finite-dimensional immersed smooth submanifold (of class $C^q$ if $f$ is of class $C^q$) and it has the following properties.

**Theorem 4.6.** *If $x$, $y \in M$, then:*

(1) $W(x) \cap W(y) = \emptyset$ *if* $y \notin W(x)$;
(2) $W(x) = W(y)$ *if* $y \in W(x)$;
(3) $f(W(x)) = W(f(x))$;
(4) $W(x)$ *can be characterized as follows: it consists of all points* $y \in M$ *for which* (4.4) *holds.*

**Remark 4.7.** For every $x \in M$ and $y \in W(x)$ we have

$$\rho(f^n(y), f^n(x)) \to 0 \text{ as } n \to \infty$$

(with an exponential rate) but $W(x)$ may not consist of *all* such points due to a possible contraction in other directions in the tangent bundle.

*Proof of the theorem.* Fix $x \in M$, pick a point $y \in W(x)$ and connect $x$ and $y$ by a curve $\gamma$ whose length does not exceed $2\rho_W(x, y)$. There are points $z_0, z_1, \dots, z_k \in \gamma$ such that $z_0 = x$, $z_k = y$ and $z_{i+1} \in V(z_i)$ for all $i = 0, \dots, k-1$. It follows from Local Stable Manifold Theorem 4.3 that for all $i$ and $n \geq 0$,

$$f^n(z_{i+1}) \in V(f^n(z_i))$$

and

$$\rho(f^n(z_{i+1}), f^n(z_i)) \leq C(\lambda + \varepsilon)^n \rho(z_{i+1}, z_i), \tag{4.39}$$

where $C > 0$ is a constant independent of $i$. Fix $n \geq 0$. Summing (4.39) over $i$ we obtain that

$$\sum_{i=0}^{k-1} \rho(f^n(z_{i+1}), f^n(z_i)) \leq C(\lambda + \varepsilon)^n \sum_{i=0}^{k-1} \rho(z_{i+1}, z_i). \tag{4.40}$$

Clearly, $\rho_W(f^n(x), f^n(y))$ does not exceed the sum on the left-hand side of (4.40). The sum on the right-hand side of (4.40) can be estimated from above by $C_1 \operatorname{length} \gamma$ where $C_1 > 0$ is a constant. This implies Statement 4. Statement 3 is obvious and Statements 1 and 2 follow form Statement 4. $\square$

**4.6. Filtrations of foliations.** As an immediate corollary of the Stable Manifold Theorem 4.1 we obtain the integrability theorem in the case when the tangent space admits a $df$-invariant splitting

$$TM = \bigoplus_{i=1}^{t} E_i \tag{4.41}$$

with

$$\lambda_i \leq \|df|E_i\| \leq \mu_i, \tag{4.42}$$

where

$$0 < \lambda_1 \leq \mu_1 < \cdots < \lambda_t \leq \mu_t. \tag{4.43}$$

**Theorem 4.8.** *The following statements hold:*

(1) *if $\mu_k < 1$ then the distribution*

$$F_k^s = \bigoplus_{i=1}^{k} E_i$$

*is integrable and the maximal integral manifolds of this distribution generate a foliation $W_k^s$ of $M$. The global leaf $W_k^s(x)$ through $x \in M$ is a $C^1$-immersed submanifold of $M$;*

(2) *if $\lambda_k > 1$ then a similar statement holds about the distribution*

$$F_k^u = \bigoplus_{i=k}^{t} E_i;$$

*the corresponding foliation is $W_k^u$ and its leaves are $W_k^u(x)$, $x \in M$;*

(3) *the foliation $W_k^s$ is $f$-invariant and contracting, i.e., for any $x \in M$, $y \in W_k^s(x)$ and $n \geq 0$,*

$$\rho_k^s(f^n(x), f^n(y)) \leq C(\lambda_k + \varepsilon)^n \rho_k^s(x, y),$$

*where $\varepsilon$ is such that $0 < \varepsilon < \min\{\lambda_{k+1} - \mu_k, 1 - \mu_k\}$, $C = C(\varepsilon) > 0$ is a constant independent of $x$, $y$ and $n$, and $\rho_k^s$ is the distance in $W_k^s(x)$ induced by the Riemannian metric;*

(4) *the foliation $W_k^u$ is $f$-invariant and contracting for $n \leq 0$;*

(5) *if $f$ is of class $C^q$ then $W_k^s(x) \in C^q$ and $W_k^u(x) \in C^q$.*

We now have two filtrations of distributions

$$F_1^s \subset F_2^s \subset \cdots \subset F_\ell^s, \qquad F_m^u \supset \cdots \supset F_t^u$$

and the corresponding filtrations of foliations

$$W_1^s \subset W_2^s \subset \cdots \subset W_\ell^s, \qquad W_m^u \supset \cdots \supset W_t^u,$$

where $\ell$ and $m$ are such that $\mu_\ell < 1$ and $\lambda_m > 1$ (note that $m = \ell + 1$ or $\ell + 2$). $W_k^s$ is called the *$k$-stable foliation* and $W_k^u$ the *$k$-unstable foliation* for $f$. For a partially hyperbolic in the narrow sense diffeomorphism $f$ the foliations $W^s = W_\ell^s$ and $W^u = W_m^u$ are called *stable* and *unstable* foliations.

For each $k$, $1 \le k < \ell$ (respectively, $m \le k < t$), the foliation $W_k^s$ *subfoliates* the foliation $W_{k+1}^s$ (respectively, $W_{k+1}^u$ subfoliates $W_k^u$). One can show that for every $x \in M$ the leaves $W_k^s(y)$ depend $C^{N_k}$ smoothly on $y \in W_{k+1}^s(x)$, where $N_k$ is the biggest number satisfying $\mu_k < \lambda_{k+1}^{N_k}$ (see [26], Theorem 6.1; compare with Statement 4 of Theorem 5.2). A similar statement holds for $W_k^u$.

**4.7. The Inclination Lemma.** Perron's approach, described formally by Theorem 4.3, is quite powerful: one can use it to construct invariant manifolds in various situations with simple (sometimes quite obvious) modifications keeping the main ingredients of the method essentially unchanged. Here we use it to construct stable and unstable manifolds for small perturbations of Anosov diffeomorphisms. This will allow us to establish structural stability of Anosov maps, see Section 4.8.

Let $f$ be a $C^q$ Anosov diffeomorphism of a smooth compact Riemannian manifold $M$ and $TM = E^s \oplus E^u$ the invariant splitting of the tangent bundle such that

$$\|df|E^s\| \le \lambda, \quad \|(df|E^u)^{-1}\|^{-1} \ge \mu,$$

with $0 < \lambda < 1 < \mu$.

Given $x \in M$, consider the family of local stable manifolds

$$\{V_{m,f} = V^s(f^m(x))\}_{m \in \mathbb{Z}}.$$

Let $T$ be a small perturbation of $f$. It is an Anosov diffeomorphism. For any point $y$ in a small neighborhood $U$ of $x$ and any $m \in \mathbb{Z}$ consider a local stable manifold $V^s(T^m(y))$ for $T$. It turns out that one can identify a unique family $\{V_{m,T}\}_{m \in \mathbb{Z}}$ of local (stable) manifolds with the property that for any $m \in \mathbb{Z}$,

$$\rho(f^m(x), V_{m,T}) \le C,$$

where $C > 0$ is a constant. More precisely, the following statement holds.

**Theorem 4.9.** *Given $\delta > 0$ there is $\varepsilon > 0$ with the following property. Let $T: M \to M$ be a $C^q$ diffeomorphism such that $d_{C^1}(f, T) \le \varepsilon$. Given $x \in M$, there is a family of local $C^q$ manifolds $\{V_{m,T}\}_{m\in\mathbb{Z}}$ such that:*

(1) *$d(f^m(x), V_{m,T}) \le \delta$;*
(2) *$V_{m,T}$ is a local stable manifold for $T$, i.e., $T(V_{m,T}) \subset V_{m+1,T}$ and, for any $y, z \in V_{m,T}$ and $n \ge 0$, $d(T^n(y), T^n(z)) \le C\kappa^n d(y,z)$, where $C > 0$ and $0 < \kappa < 1$ are constants;*
(3) *the family $\{V_{m,T}\}$ is unique, i.e., any other family satisfying Statements 1 and 2 coincides with $\{V_{m,T}\}$.*

*Proof.* Given $x \in M$ we find a family of local manifolds at $x$ in the following form: for every $m \in \mathbb{Z}$,

$$V_{m,T} = \exp_x\{(v, \psi_m(v)) : v \in B^s(r_0)\}, \tag{4.44}$$

where $B^s(r_0)$ is the ball in $E^s(x)$ centered at the origin of radius $r_0$ and $\psi_m : B^s(r_0) \to F(x)$ is a smooth map satisfying $\|\psi_m(0)\| \le \varepsilon$ and $\|d\psi_m(0)\| \le \varepsilon$. The number $r_0$ is sufficiently small and is independent of $x$.

To construct the family of functions $\{\psi_m\}_{m\in\mathbb{Z}}$ consider the local map

$$\widetilde{T}_x = \exp_{f(x)}^{-1} \circ T \circ \exp_x : B^s(r_0) \times B^u(r_0) \to T_{f(x)}M, \tag{4.45}$$

where $B^u(r_0) \subset E^u(x)$ is the ball of radius $r_0 > 0$ centered at the origin. The reader should notice that the action of the perturbation $T$ is evaluated along the trajectory of $x$ under the map $f$ (compare (4.45) with (4.6)).

Using the coordinate system associated with the splitting $TM = E^s \oplus E^u$, which is invariant under $df$ but not under $dT$, one can write for $v \in B^s(r_0)$ and $w \in B^u(r_0)$,

$$\widetilde{T}_x(v,w) = (A(x)v + C(x)w + g_x(v,w),\ D(x)v + B(x)w + h_x(v,w)).$$

Here

$$A(x) : E^s(x) \to E^s(f(x)), \quad B(x) : E^u(x) \to E^u(f(x))$$

are linear maps satisfying

$$\|A(x)\| \le \lambda + \varepsilon, \quad \|B(x)^{-1}\|^{-1} \ge \mu - \varepsilon \tag{4.46}$$

with

$$0 < \lambda < \lambda + \varepsilon < 1 < \mu - \varepsilon < \mu. \tag{4.47}$$

Furthermore,

$$C(x) : E^u(x) \to E^s(f(x)), \quad D(x) : E^s(x) \to E^u(f(x))$$

are linear maps such that

$$\|C(x)\| \le \varepsilon, \quad \|D(x)\| \le \varepsilon. \tag{4.48}$$

Finally,

$$g_x : B^s(r_0) \times B^s(r_0) \to E^s(f(x)), \quad h_x : B^s(r_0) \times B^u(r_0) \to E^u(f(x))$$

are non-linear maps satisfying

$$\|g_x(0,0)\| \leq \varepsilon, \quad \|h_x(0,0)\| \leq \varepsilon, \tag{4.49}$$

$$dg_x(0,0) = 0, \qquad dh_x(0,0) = 0, \tag{4.50}$$

$$\|dg_x(v,w)\| \leq C_1, \quad \|dh_x(v,w)\| \leq C_1, \tag{4.51}$$

where $C_1 > 0$ is a constant. Using trivializations $\tau_m : T_{f^m(x)}M \to \mathbb{R}^p$, where $p = \dim M$, we obtain, as in (4.13) and (4.14), the sequence of maps

$$G_m = \tau_{m+1} \circ \widetilde{T}_{f^m(x)} \circ \tau_m^{-1} \colon B^\ell(r_0) \times B^{p-\ell}(r_0) \to \mathbb{R}^p \quad (\ell = \dim E(x))$$

such that

$$G_m(v,w) = (A_m v + \widetilde{g}_m(v,w), B_m w + \widetilde{h}_m(v,w)).$$

Here

$$A_m \colon \mathbb{R}^\ell \to \mathbb{R}^\ell, \quad B_m \colon \mathbb{R}^{p-\ell} \to \mathbb{R}^{p-\ell}$$

are linear maps and by (4.46),

$$\|A_m\| \leq \lambda + \varepsilon, \quad \|B_m{}^{-1}\|^{-1} \geq \mu - \varepsilon. \tag{4.52}$$

Furthermore,

$$\widetilde{g}_m \colon B^\ell(r_0) \times B^{p-\ell}(r_0) \to \mathbb{R}^\ell, \quad \widetilde{h}_m \colon B^\ell(r_0) \times B^{p-\ell}(r_0) \to \mathbb{R}^{p-\ell}$$

are nonlinear maps and by (4.48)–(4.51),

$$\|\widetilde{g}_m(0,0)\| \leq \varepsilon, \quad \|\widetilde{h}_m(0,0)\| \leq \varepsilon, \tag{4.53}$$

$$\|d\widetilde{g}_m(0,0)\| \leq \varepsilon, \quad \|d\widetilde{h}_m(0,0)\| \leq \varepsilon, \tag{4.54}$$

$$\|d\widetilde{g}_m(v,w)\| \leq C_2, \quad \|d\widetilde{h}_m(v,w)\| \leq C_2, \tag{4.55}$$

where $C_2 > 0$ is a constant.

Conditions (4.53) and (4.54) are more general than similar Conditions (4.9) and (4.10) and this requires appropriate modifications of the proof.

Since the family of local stable manifolds $\{V_{m,T}\}_{m\in\mathbb{Z}}$ is invariant under $T$ we have for every $m$,

$$T(V_{m,T}) \subset V_{m+1,T}.$$

This yields the following equation for the functions $\psi_m$:

$$B_m\psi_m(v) + \widetilde{h}_m(v, \psi_m(v)) = \psi_{m+1}(A_m v + \widetilde{g}_m(v, \psi_m(v))). \tag{4.56}$$

To solve this equation we proceed as in the proof of Theorem 4.1, using instead of Local Stable Manifold Theorem 4.3 the following statement, known as the *Inclination Lemma* (or λ-Lemma; see [29], Section 6.2).

**Lemma 4.10.** *Given $\delta > 0$, there is $\varepsilon > 0$ such that for any $C^q$ diffeomorphism $T$ with $d_{C^1}(f, T) \le \varepsilon$ the following holds. There are constants $D > 0$, $0 < r < r_0$ and a map $\psi : B^\ell(r) \to \mathbb{R}^{p-\ell}$ satisfying:*

(1) *$\psi$ is of class $C^q$ and $\|\psi(0)\| \le \delta$, $\|d\psi(0)\| \le \delta$;*
(2) *for any $m \ge 0$ and $v \in B^\ell(r)$,*

$$\left(\prod_{i=0}^{m-1} G_i\right)(v, \psi(v)) \in B^\ell(r_0) \times B^{p-\ell}(r_0),$$

$$\left\|\left(\prod_{i=0}^{m-1} G_i\right)(v, \psi(v))\right\| \le D(\lambda + \varepsilon)^m \|(v, \psi(v))\|;$$

(3) *if $(v, w) \in B^\ell(r) \times B^{p-\ell}(r_0)$ is such that*

$$\left(\prod_{i=0}^{m-1} G_i\right)(v, w) \in B^\ell(r_0) \times B^{p-\ell}(r_0), \qquad \left\|\left(\prod_{i=0}^{m-1} G_i\right)(v, w)\right\| \le K(\lambda + \varepsilon)^m$$

*for some $K > 0$ and every $m \ge 0$, then $w = \psi(v)$.*

*Proof.* We shall follow with some modifications the proof of Theorem 4.3. Consider the linear space $\Gamma$ of sequences of vectors $z = \{z(m) \in \mathbb{R}^p\}_{m \ge 0}$ satisfying

$$\|z\|_\Gamma = \sup_{m \ge 0} \|z(m)\| < \infty$$

(this corresponds to (4.19) with $\kappa = 1$). $\Gamma$ is a Banach space with the norm $\|z\|_\Gamma$. Consider the open subset $W \subset \Gamma$ and the operator $\Phi \colon B^\ell(r_0) \times W \to \Gamma$ as in (4.20), (4.21), (4.22) and (4.23) (with $g$ and $h$ replaced by $\widetilde{g}$ and $\widetilde{h}$).

Using (4.55) and (4.53), we obtain (compare with (4.24) and (4.25))

$$\|\widetilde{g}_n(z)\| \le C_2\|z\| + \varepsilon, \quad \|\widetilde{h}_n(z)\| \le C_2\|z\| + \varepsilon. \tag{4.57}$$

In view of (4.47), (4.52) and (4.57), we have the following estimate

$$
\begin{aligned}
\|\Phi(y,z)\|_\Gamma &= \sup_{m\geq 0}\|\Phi(y,z)(m)\| \leq \sup_{m\geq 0}\|z(m)\| \\
&+ \sup_{m\geq 0}\left\{\left(\prod_{i=0}^{m-1}\|A_i\|\right)\|y\| + \sum_{n=0}^{m-1}\left(\prod_{i=n+1}^{m-1}\|A_i\|\right)(C_2\|z(n)\|+\varepsilon)\right. \\
&\left.+\sum_{n=0}^{\infty}\left(\prod_{i=0}^{n}\|B_{i+m}^{-1}\|\right)(C_2\|z(n+m)\|+\varepsilon)\right\} \\
&\leq \|z\|_\Gamma + \sup_{m\geq 0}\lambda^m\|y\| \\
&+ (C_2\|z\|_\Gamma+\varepsilon)\sup_{m\geq 0}\left[\sum_{n=0}^{m-1}\lambda^{m-n-1}+\sum_{n=0}^{\infty}\mu^{-(n+1)}\right].
\end{aligned}
$$

It follows that

$$\|\Phi(y,z)\|_\Gamma \leq \|y\| + C_3\|z\|_\Gamma + C_4\varepsilon,$$

where $C_3 > 0$ and $C_4 > 0$ are constants. This implies that the map $\Phi$ is well-defined. We also have that $\Phi(0,0) = \Phi_0$ for some (in general, not zero) $\Phi_0$ with $\|\Phi_0\|_\Gamma \leq \varepsilon$. We leave it to the reader to show that the map $\Phi$ is of class $C^1$ with partial derivatives given by (4.26), (4.27) and (4.28) (where $g$ and $h$ should be replaced by $\widetilde{g}$ and $\widetilde{h}$). Although $\mathcal{A}(0)$ may not be zero in view of (4.54) we have that $\|\mathcal{A}(0)\|_\Gamma \leq \varepsilon$. One can further verify that the map $\Phi$ is $q$ times differentiable with derivatives of higher order given by (4.37) (where $g$ and $h$ should be replaced again by $\widetilde{g}$ and $\widetilde{h}$).

Applying the Implicit Function Theorem as in the proof of Theorem 4.3 yields that, for sufficiently small $\varepsilon$, there exist a number $0 < r \leq r_0$ and a map $\varphi\colon B^\ell(r) \to W$ of class $C^q$ satisfying (compare with (4.29))

(1) $\Phi(y,\varphi(y)) = \Phi_0$ for all $y \in B^\ell(r)$;
(2) $\varphi(0) = \varphi_0$, $d\varphi(0) = \varphi_1$ with $\|\varphi_0\|_\Gamma \leq \delta$ and $\|\varphi_1\|_\Gamma \leq \delta$.

The desired map $\psi$ is given by (4.34). □

Observe that for each $m$ the local manifold $V_{m,T}$ (see (4.44)) is stable with respect to $T$ and smooth of class $C^q$. In particular, $T_yV_{m,T} = E_T^s(y)$ for every $y \in V_{m,T}$. This completes the proof of the theorem. □

**4.8. Structural stability of Anosov diffeomorphisms.** As a manifestation of Theorem 4.9 we obtain the following Structural Stability theorem for Anosov diffeomorphisms.

**Theorem 4.11** (Anosov, [1]). *There is $\delta_0 > 0$ such that for every $0 < \delta \leq \delta_0$ one can find $\varepsilon > 0$ for which the following holds: if $g : M \to M$ is a $C^q$*

*diffeomorphism with $d_{C^1}(f,g) \le \varepsilon$ then $f$ and $g$ are topologically conjugate via a homeomorphism $h$ for which $d_{C^0}(h, Id) \le \delta$.*

*Sketch of the Proof.* If $\varepsilon$ is sufficiently small, any diffeomorphism $g$, for which $d_{C^1}(f,g) \le \varepsilon$, is Anosov. Applying Theorem 4.9 to $f$, given $x \in M$, one can construct a family of local stable manifolds, $\{V^s_{m,g}\}_{m\in\mathbb{Z}}$ such that, if $\varepsilon$ is sufficiently small, then for any $m \in \mathbb{Z}$,

$$\rho(f^m(x), V^s_{m,g}) \le \delta.$$

Applying then Theorem 4.9 to $f^{-1}$, given $x \in M$, one can construct a family of local unstable manifolds, $\{V^u_{m,g}\}_{m\in\mathbb{Z}}$ such that, if $\varepsilon$ is sufficiently small, then for any $m \in \mathbb{Z}$,

$$\rho(f^m(x), V^u_{m,g}) \le \delta.$$

For sufficiently small $\delta$ the intersection $V^s_{0,g} \cap V^u_{0,g}$ consists of a single point $y$ and the map $h(x) = y$ can be shown to be a homeomorphism of $M$ which gives the desired conjugacy between $f$ and $g$. □

There is another somewhat more powerful approach to structural stability of Anosov maps based on the concept of pseudo orbits. A sequence of points $\{x_n\}_{n\in\mathbb{Z}}$ is called an *$\varepsilon$-pseudo orbit* (or simply a *pseudo orbit*) if $\rho(f(x_n), x_{n+1}) \le \varepsilon$ for every $n \in \mathbb{Z}$. Roughly speaking this means that starting from the point $x_0$ we apply $f$ to obtain $f(x_0)$, budge it by $\le \varepsilon$ to produce $x_1$, apply $f$ to $x_1$ to obtain $f(x_1)$, budge it by $\le \varepsilon$ to produce $x_2$, etc.

We say that an orbit $\{f^n(y)\}_{n\in\mathbb{Z}}$ *$\delta$-shadows* (or simply *shadows*) a pseudo orbit $\{x_n\}_{n\in\mathbb{Z}}$ if $\rho(x_n, f^n(y)) \le \delta$ for all $n \in \mathbb{Z}$. The Shadowing theorem for Anosov diffeomorphisms (see [29], Theorem 18.1.2) claims that *any pseudo orbit is shadowed by a unique orbit.* The Structural Stability theorem is a corollary of this statement. Indeed, if a diffeomorphism $g$ is a perturbation of $f$, so that $d_{C^1}(f,g) \le \varepsilon$, then for any $x \in M$ the sequence of points $\{x_n = g^n(x)\}_{n\in\mathbb{Z}}$ is an $\varepsilon$-pseudo orbit for $f$ (to see this write $\rho(f(x_n), x_{n+1}) = \rho(f(x_n), g(x_n)) \le \varepsilon$). Thus, it is shadowed by a unique orbit $\{f^n(y)\}_{n\in\mathbb{Z}}$ and the map $h\colon x \to y$ can be shown to be a homeomorphism which conjugates $f$ and $g$ (see [29], Section 18.1).

There is a version of the notions of pseudo orbits and shadowing for partially hyperbolic systems which will be discussed in Section 5.5.

# 5. Central Foliations

**5.1. Normal hyperbolicity.** The notion of normal hyperbolicity in dynamical systems was introduced by Hirsch, Pugh and Shub in [25] (see also [26]; a particular case of normal hyperbolicity was considered by R. Sacker, [45] in his work on partial differential equations). Normal hyperbolicity and partial hyperbolicity theories are closely related in their results and methods. Moreover, the former provides techniques to study integrability of the central distribution and robustness of the central foliation for partially hyperbolic systems (see Sections 5.4 and 5.5).

Let $f$ be a diffeomorphism of class $C^q$, $q \geq 1$, of a compact smooth Riemannian manifold $M$ which is connected and without boundary. Let also $N = N_f$ be a compact smooth submanifold of $M$ which is invariant under $f$, i.e., $f(N) = N$. The map $f$ is called *normally hyperbolic to* $N$ if $f$ is partially hyperbolic on $N$ in the narrow sense. More precisely, for every $x \in N$ there is an invariant splitting

$$
\begin{gathered}
T_xM = E^s(x) \oplus T_xN \oplus E^u(x), \\
df E^s(x) = E^s(f(x)), \quad df E^u(x) = E^u(f(x))
\end{gathered}
\tag{5.1}
$$

such that

$$
\begin{gathered}
\lambda_1 \leq \|df|E^s(x)\| \leq \mu_1, \quad \lambda_2 \leq \|df|T_xN\| \leq \mu_2, \\
\lambda_3 \leq \|df|E^u(x)\| \leq \mu_3,
\end{gathered}
\tag{5.2}
$$

where the numbers $\lambda_i$ and $\mu_i$, $i = 1, 2, 3$, satisfy (2.6). It follows from Proposition 3.9 that the splitting (5.1) is Hölder continuous.

By the Local Stable Manifold Theorem 4.3, one can construct, for every $x \in N$, local stable and unstable manifolds at $x$, $V^s(x)$ and $V^u(x)$ respectively, such that

(1) $x \in V^s(x)$, $x \in V^u(x)$;
(2) $T_xV^s(x) = E^s(x)$, $T_xV^u(x) = E^u(x)$;
(3) for $n \geq 0$,

$$\rho(f^n(x), f^n(y)) \leq C(\mu_1 + \varepsilon)^n \rho(x, y), \quad y \in V^s(x),$$

$$\rho(f^{-n}(x), f^{-n}(y)) \leq C(\lambda_3 - \varepsilon)^n \rho(x, y), \quad y \in V^u(x),$$

where $C > 0$ is a constant and $\varepsilon > 0$ is sufficiently small.

Set

$$
V^{so}(N) = \bigcup_{x \in N} V^s(x), \quad V^{uo}(N) = \bigcup_{x \in N} V^u(x).
\tag{5.3}
$$

These are topological manifolds called local *stable* and *unstable* manifolds of $N$. They are $f$-invariant and

$$N = V^{so}(N) \cap V^{uo}(N).$$

**Theorem 5.1** (Hirsch, Pugh and Shub, [26], Proposition 5.7). *$V^{uo}(N)$ and $V^{so}(N)$ are Lipschitz submanifolds in $M$.*

*Proof.* We shall show that $V^{uo}(N)$ is a Lipschitz submanifold in $M$. The proof for $V^{so}(N)$ is similar. First, we introduce a special tubular neighborhood of $N$ and a special coordinate system in it.

Given small $\varepsilon > 0$, choose smooth distributions $\widetilde{E}^s, \widetilde{E}^u \subset TM$ which approximate invariant distributions $E^s$ and $E^u$, i.e.,

$$\angle(\widetilde{E}^s(p), E^s(p)) \le \varepsilon, \quad \angle(\widetilde{E}^u(p), E^u(p)) \le \varepsilon \tag{5.4}$$

for every $p \in N$ (without loss of generality we may assume that $\widetilde{E}^s$ and $\widetilde{E}^u$ are of $C^\infty$ class of smoothness). For small $r_0 > 0$ let $\widetilde{B}^s(p, r_0) \subset \widetilde{E}^s(p)$ and $\widetilde{B}^u(p, r_0) \subset \widetilde{E}^u(p)$ be balls centered at zero of radius $r$). One can arrange the sets

$$V(p) = \exp_p(\widetilde{B}^s(p, r_0) \times \widetilde{B}^u(p, r_0)), \tag{5.5}$$

corresponding to different $p \in N$, to be disjoint. Thus, the set

$$U_{r_0}(N) = \bigcup_{p \in N} V(p) \tag{5.6}$$

is a tubular neighborhood of $N$. Up to the identification by the exponential map every point $z \in U_{r_0}(N)$ has coordinates $z = (p, v, w)$, where $p \in N$, $v \in \widetilde{E}^s(p)$ and $w \in \widetilde{E}^u(p)$. Let

$$X = \bigcup_{p \in N} \widetilde{B}^u(p, r_0).$$

We can view $V^{uo}(N)$ as the graph of a map $\psi\colon X \to \widetilde{E}^s$, and we wish to show that $\psi$ is a Lipschitz map. To this end choose $\delta > 0$ and two points $z, z' \in V^{uo}(N)$ such that $\rho(z, z') \le \delta$. Write $z = (x, y)$, $z' = (x', y')$ with $y = \psi(x)$ and $y' = \psi(x')$. It suffices to show that for some sufficiently small $\varepsilon$, any $\delta > 0$ and any $z$ and $z'$ as above,

$$\|y - y'\| \le \|x - x'\|. \tag{5.7}$$

Assuming the contrary, one can find, for some small $\varepsilon > 0$ and some $p, p' \in N$, two points $z \in B^u(p, \varepsilon)$ (the $\varepsilon$-ball in $V^u(p)$ centered at $p$ of radius $\varepsilon$) and $z' \in B^u(p', \varepsilon)$ which are very close to each other and whose distance apart in the unstable direction is greater or equal to their distance apart in the central direction (see Figure 7). Apply $f^{-n}$ to these points and observe that the distance apart in the central direction cannot expand as fast as the distance apart in the unstable direction must expand (by normal hyperbolicity). Likewise $f^{-n}(z)$, $f^{-n}(z')$ must lie very near $N$ and $f^{-n}(p)$, $f^{-n}(p')$ must not be very far apart.

This is incompatible with the local unstable manifolds $V^u(p)$ being uniformly tangent to $E^u(p)$, $p \in N$.

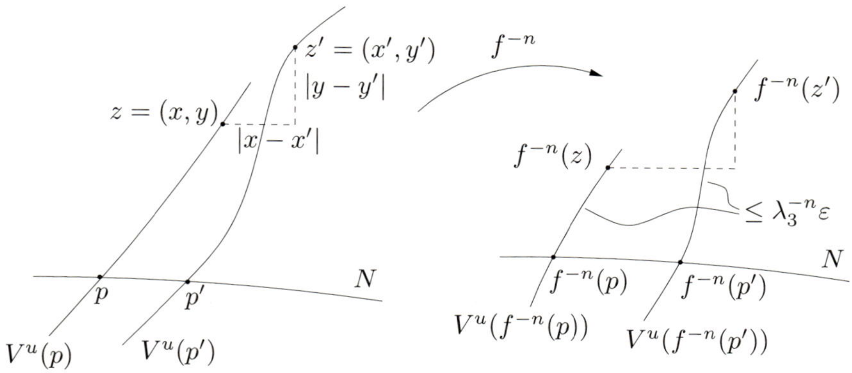


FIGURE 7. To the proof of Theorem 5.1.

Since such $z$, $z'$ cannot occur, $V^{uo}(N)$ is a Lipschitz submanifold. □

Once $V^{uo}(N)$ and $V^{uo}(N)$ are proven to be Lipschitz submanifolds, one can use results in [26] (see Theorem 3.5) to conclude that they are indeed smooth.

**5.2. Local stability of normally hyperbolic manifolds.** In [26], Hirsch, Pugh and Shub used Hadamard's method for constructing local stable and unstable manifolds through $N$. Their approach does not rely on the existence of local stable manifolds through individual points $x \in N$ but instead, builds local stable and unstable manifolds through $N$ as a whole. Of course, *a posteriori* one can derive the formula (5.3). Hirsch, Pugh and Shub obtained a more complete information on local stable and unstable manifolds through $N$. In particular, they showed that a normally hyperbolic manifold $N$ survives under small perturbation of the system, thus establishing stability of normal hyperbolicity. We state here their results without proof referring to Sections 4–6 of their paper.

Define the numbers $\ell_s, \ell_u \le q$ to be the biggest integers such that

$$\mu_1 < \lambda_2^{\ell_u}, \quad \mu_2^{\ell_s} < \lambda_3, \quad \ell = \min\{\ell_s, \ell_u\} \tag{5.8}$$

**Theorem 5.2.** *Let $f$ be a diffeomorphism of class $C^q$, $q \geq 1$ which is normally hyperbolic to a compact smooth manifold $N$. Then the following statements hold.*

(1) *Existence: there exist locally $f$-invariant submanifolds $V^{so}(N)$ and $V^{uo}(N)$ tangent to $E^s \oplus TN$ and $E^u \oplus TN$ respectively.*
(2) *Uniqueness: if $N'$ is an $f$-invariant set which lies in an $\varepsilon$-neighborhood $U_\varepsilon(N)$ of $N$, for sufficiently small $\varepsilon$, then $N' \subset V^{so}(N) \cup V^{uo}(N)$.*
(3) *Characterization: $V^{so}(N)$ (respectively, $V^{uo}(N)$) consists of all points $y$ for which $\rho(f^n(y), N) \leq r$ for all $n \geq 0$ (respectively, $n \leq 0$) and some small $r > 0$; indeed, $\rho(f^n(y), N) \to 0$ with an exponential rate as $n \to +\infty$ (respectively, as $n \to -\infty$).*
(4) *Smoothness: $V^{so}(N)$ and $V^{uo}(N)$ are submanifolds in $M$ of class $C^{\ell_s}$ and $C^{\ell_u}$ respectively; in particular, $N$ is a $C^\ell$ submanifold.*
(5) *Lamination: $V^{so}(N)$ and $V^{uo}(N)$ are fibered by $V^s(x)$ and $V^u(x)$, $x \in N$; see (5.3).*

**Theorem 5.3.** *Let $f$ be a diffeomorphism of class $C^q$, $q \geq 1$, which is normally hyperbolic to a compact smooth manifold $N$. Then for every $\delta > 0$ there exists $r > 0$ and $\varepsilon > 0$ such that the following statements hold:*

(1) *for every diffeomorphism $T$ of class $C^q$ with $d_{C^1}(f, T) \leq \varepsilon$, there exists a smooth submanifold $N_T$, invariant under $T$, to which $T$ is normally hyperbolic; $N_T$ lies in an $r$-neighborhood $U_r(N_f)$ of $N_f$.*
(2) *$V_T^{so}(N) \in C^{\ell_s}$, $V_T^{uo}(N) \in C^{\ell_u}$ and $N_T \in C^\ell$ (where the numbers $\ell_s$, $\ell_u$, and $\ell$ are given by (5.8)); they depend continuously on $T$ in the $C^1$ topology;*
(3) *there exists a homeomorphism $H\colon U_r \to M$ which is $\delta$-close to the identity map in the $C^0$ topology and such that $H(N_f) = N_g$.*

**5.3. Integrability of the central foliation.** Let $M$ be a compact smooth Riemannian manifold and $f\colon M \to M$ a diffeomorphism which is partially hyperbolic in the narrow sense with a $df$-invariant splitting of the tangent bundle (2.7) satisfying (2.8). The central distribution $E^c$ may not, in general, be integrable as Example 6.1 below illustrates. Note that non-integrability is an open property (see Theorem 5.16).

We describe some conditions which guarantee integrability of the central distribution. In fact, these conditions guarantee a stronger property of unique integrability which we introduce now.

Let $E$ be a continuous $k$-dimensional distribution on $M$. It is called

(1) *uniquely integrable* if there is a foliation $W$ with $k$-dimensional leaves such that any $C^1$ curve $\sigma\colon \mathbb{R} \to M$ satisfying $\dot\sigma(t) \in E(\sigma(t))$ for all

$t$, is contained in $W(\sigma(0))$ (in particular, it follows that $T_xW(x) = E(x)$ for any $x \in M$);

(2) *local uniquely integrable* if for each $x \in M$ there is a $k$-dimensional smooth submanifold $W_{\mathrm{loc}}(x)$ and $\alpha(x) > 0$ such the following holds: let $\sigma\colon [0,1] \to M$ be a piecewise $C^1$ curve satisfying a) $\sigma(0) = x$; b) $\dot\sigma(t) \in E(\sigma(t))$ for $t \in [0,1]$; c) $\operatorname{length}\sigma < \alpha(x)$; then $\sigma$ is contained in $W_{\mathrm{loc}}(x)$.

Obviously, if $E$ is locally uniquely integrable then it is integrable and the integrable foliation is unique.

A foliation $W$ of $M$ is said to be *quasi-isometric* if there are $a > 0$ and $b > 0$ such that for every $x \in M$ and every $y \in W(x)$,

$$\rho_W(x,y) \le a \cdot \rho(x,y) + b,$$

(recall that $\rho_W$ is the distance along the leaves of the foliation $W$). Denote by $\widetilde{M}$ the universal cover of $M$ and by $\widetilde{W}^s$ and $\widetilde{W}^u$ the lifts of the stable and unstable foliations to $\widetilde{M}$.

**Theorem 5.4** (Brin, [8]). *If $\widetilde{W}^s$ and $\widetilde{W}^u$ are quasi-isometric in the universal cover $\widetilde{M}$, then the distributions $E^{cs}$, $E^{cu}$ and $E^c$ are locally uniquely integrable.*

*Proof.* We will prove the integrability of $E^{cs}$. The integrability of $E^{cu}$ follows by reversing the time. The integrability of $E^c$ follows by observing that every $C^1$ curve starting at $x \in M$ and tangent to $E^c$ is also tangent to $E^{cs}$ and $E^{cu}$ and therefore is contained in $W^{cs}(x) \cap W^{cu}(x)$.

**Lemma 5.5.** *Suppose $W$ is a $k$-dimensional foliation of $M$ and $E$ an $\ell$-dimensional distribution ($k + \ell = m$) which is (uniformly) transverse to $W$, i.e., $T_xW(x) \cap E(x) = \{0\}$ for each $x \in M$.*

*Then there is a constant $C_0 > 0$ and for every $\delta > 0$ there is $\varepsilon_0 > 0$ such that the following holds: for any $x, y \in M$ with $\rho(x,y) < \varepsilon_0$ one can find a $C^1$ curve $\sigma : [0,1] \to M$ satisfying:*

(1) *$\sigma(0) = x$, $\sigma(1) \in B_W(y,\delta)$ (the $\delta$-ball in $W(y)$ centered at $y$),*
(2) *$\dot\sigma(t) \in E(\sigma(t))$ for $t \in [0,1]$,*
(3) *$\operatorname{length}(\sigma) \le C_0\rho(x,y)$, and $\rho_W(y,\sigma(1)) \le C_0\rho(x,y)$.*

*Proof of the lemma.* By choosing $\varepsilon_0$ small enough we may assume that a neighborhood of $x$ containing $y$ is identified by a coordinate chart (with uniformly bounded derivative) with the unit ball in $\mathbb{R}^m$. For small $\delta > 0$ the ball $B = B_W(y,\delta) \subset W$ can be identified with a $k$-dimensional $C^1$ submanifold in $\mathbb{R}^m$. There is a point $z \in B$ for which $\rho(x,z) = \rho(x,B)$ and

$\rho_W(z, y) \leq C\rho(x, y)$, where $C$ depends only on $W$. In particular, $\rho(x, z) \leq \rho(x, y)$ and $x - z \perp T_x B$. Consider the union

$$S = \bigcup_{0 \leq t \leq 1} (B + (1-t)(x-z))$$

of the parallel translates of $B$ along the segment connecting $x$ to $z$. It is a $(k+1)$-dimensional $C^1$ surface which is a Riemannian product of the interval $[0, \|x - z\|]$ and $B$ (see Figure 8). We refer to $B$ and $x - z$ as the horizontal and vertical components, respectively. For $s \in S$ let $L(s) = E(s) \cap T_s S$. Since $E$ is uniformly transverse to $B$, for sufficiently small $\varepsilon_0$ and $\delta$ the line field $L$ is a continuous on $S$ and is uniformly (in $x$, $y$ and $s$) transverse to the codimension 1 (in $S$) submanifolds $B + (1-t)(x-z)$. Therefore, for each $s \in S$ there is a unique vector $\mathbf{v}(s) \in L(s)$ whose vertical component is $z - x$. The uniform transversality of $E$ and $B + (1-t)(x-z)$ implies that $\mathbf{v}$ is a uniformly (in $x$, $y$ and $s$) bounded continuous vector field on $S$. For each $s \in S$ the horizontal component of $\mathbf{v}(s)$ is bounded by a constant $C'$ which depends on $\varepsilon_0$ and $\delta$ but not on $x$, $y$ or $s$. By the existence theorem for ordinary differential equations, there is a (possibly not unique) solution $\sigma(\cdot)$ of $\dot{s} = \mathbf{v}(s)$ with $\sigma(0) = x$. The solution stays within the $C'$-cone near the vertical segment in $S$. Therefore, if $\varepsilon_0$ and $\delta$ are small enough, the interval of definition of the solution can be extended to $[0, 1]$ so that $\sigma(1) \in B$,

$$\text{length}\,(\sigma) \leq \sqrt{1 + C'^2}\rho(x, z) \leq C_0 \rho(x, y)$$

and $\rho_W(y, \sigma(1)) \leq C_0 \rho(x, y)$. □

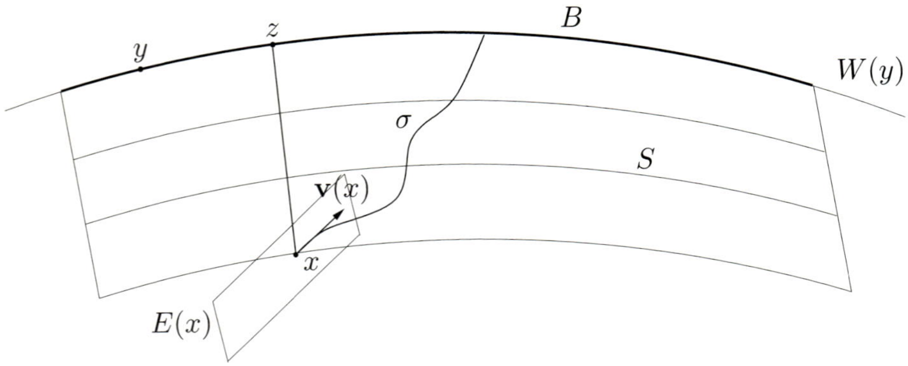


FIGURE 8. To the proof of Lemma 5.6.

To prove integrability of $E^{cs}$ let us lift all distributions and foliations to the universal cover $\widetilde{M}$. Fix $r \in (0, \varepsilon_0)$. For $x \in \widetilde{M}$ denote by $W_r^{cs}(x)$ the set of ends of all piecewise $C^1$ curves $\sigma : [0,1] \to M$ such that $\sigma(0) = x$, $\dot{\sigma}(t) \in E^{cs}(\sigma(t))$, and length$(\sigma) < r$.

**Lemma 5.6.** *Let $y, z \in \widetilde{M}$ be such that $y, z \in W_r^{cs}(x)$ and $z \in W^u(y)$. Then $z = y$.*

*Proof of the lemma.* By assumption, there are $C^1$ curves $\sigma_y$ and $\sigma_z$ tangent to $E^{cs}$ and connecting $x$ to $y$ and $z$, respectively. Consider the points $f^n(x)$, $f^n(y)$ and $f^n(z)$. It follows from the definition of partial hyperbolicity that

$$\text{length}\,(f^n(\sigma_y)) \le \mu_2^n \cdot \text{length}\,(\sigma_y), \quad \text{length}\,(f^n(\sigma_z)) \le \mu_2^n \cdot \text{length}\,(\sigma_z),$$

and therefore,

$$\rho(f^n(y), f^n(z)) \le \mu_2^n\,(\text{length}\,(\sigma_y) + \text{length}\,(\sigma_z)).$$

On the other hand, let $\sigma_n$ be a $C^1$ curve in $W^u(f^n(y)) = W^u(f^n(z))$ which realizes $\rho_{W^u}(f^n(y), f^n(z))$. Then

$$\rho_{W^u}(f^n(y), f^n(z)) = \text{length}\,(\sigma_n) \ge \mu_3^n \cdot \text{length}\,(f^{-n}(\sigma_n)) \ge \mu_3^n\, \rho_{W^u}(y,z).$$

Since $W^u$ is a quasi-isometric foliation,

$$\rho(f^n(y), f^n(z)) \ge (\rho_{W^u}(f^n(y), f^n(z)) - b)a \ge (\mu_3^n\, \rho_{W^u}(y,z) - b)a.$$

Note that

$$\rho(f^n(x), f^n(y)) \le \text{length}\,(f^n(\sigma_y)), \quad \rho(f^n(x), f^n(z)) \le \text{length}\,(f^n(\sigma_z)).$$

Since $\mu_2 < \mu_3$, we obtain a contradiction with the triangle inequality for $f^n(x)$, $f^n(y)$, $f^n(z)$ (with $n$ sufficiently large) unless $\rho_{W^u}(y,z) = 0$. □

We proceed with the proof of the theorem and show that $W_r^{cs}(x)$ is a $C^1$ submanifold of the same dimension as $E^{cs}$. Recall that $B^{cs}(x,r)$ and $B^u(x,r)$ denote the balls in $E^{cs}(x)$ and $E^u(x)$ respectively, centered at zero of radius $r$. Fix small $\delta > 0$. Since $W^u(y)$, $y \in \widetilde{M}$, is transverse to $E^{cs}(y)$ at $y$, for small enough $0 < \varepsilon \ll \delta$ and each $z \in B^{cs}(y,\varepsilon)$ the leaf $W^u(z)$ is transverse to $E^{cs}(y)$ at $z$. By Lemma 5.5, if $r$ and $\varepsilon < r$ are small enough, then each ball $B^u(z,\delta)$ intersects $B^{cs}(x,r)$ and by Lemma 5.6, the intersection consists of a single point $\pi(z)$. Let $z_1, z_2 \in B^{cs}(y,\varepsilon)$. Since $B^u(z,\delta)$ depends continuously on $z$ as a $C^1$ submanifold,

$$\rho(\pi(z_1), B^u(z_2,\delta)) \to 0 \text{ as } \rho(z_1,z_2) \to 0.$$

Therefore, by Lemma 5.5,

$$\rho(\pi(z_1), \pi(z_2)) \to 0 \text{ as } \rho(z_1,z_2) \to 0.$$

It follows that $\pi$ is continuous and a homeomorphism onto its image which is a neighborhood of $y = \pi(y)$ in $B^{cs}(x,r)$. It follows that $B^{cs}(x,r)$ is a topological submanifold. Obviously $E^{cs}(y)$ is the tangent plane to $B^{cs}(y,\varepsilon)$

at $y$. Since the distribution $E^{cs}$ is continuous, $B^{cs}(r,x)$ is a $C^1$ submanifold. Hence, $E^{cs}$ is integrable and the leaf $W^{cs}(x)$ of the integral foliation $W^{cs}$ passing through $x \in M$ is the set of ends of piecewise $C^1$ curves starting at $x$ and tangent to $E^{cs}$. This completes the proof of the theorem. □

We present another criterion for integrability of the central distribution. A partially hyperbolic diffeomorphism $f$ is called *center-isometric* if it acts isometrically in the central direction, i.e., $\|df(x)v\| = \|v\|$ for every $x \in M$ and $v \in E^c(x)$.

**Theorem 5.7** (Brin, [8]). *Assume that a partially hyperbolic diffeomorphism $f$ is center-isometric. Then the central distribution $E^c$ is locally uniquely integrable.*

*Proof.* We begin with the following lemma which is similar to Lemma 5.6.

**Lemma 5.8.** *There are $r_0 > 0$ and $\delta_0 > 0$ such that for every $r \in (0, r_0]$ and $\delta \in (0, \delta_0]$ the following holds: if $y, z \in W_r^{cs}(x)$ and $z \in B^u(y,r)$ then $z = y$.*

*Proof of the lemma.* Since the leaves $W^u(x)$ are $C^1$ and depend continuously on $x \in M$, there is $\delta_0 > 0$ and $c_1 > 0$ such that for any $z' \in W^u_{\delta_0}(y')$,

$$\rho_{W^u}(y', z') \le c_1 \rho(y', z').$$

By the uniform extension of $W^u$,

$$\rho_{W^u}(f^n(y), f^n(z)) \ge c_2 \mu^n \rho_{W^u}(y, z),$$

where $c_2 > 0$ and $\mu > 1$ are constants. Assuming that $z \ne y$, choose $n > 0$ so that $\rho_{W^u}(f^k(y), f^k(z)) < \delta_0$ for $k = 0, 1, \dots, n$ and $\rho_{W^u}(f^{n+1}(y), f^{n+1}(z)) \ge \delta_0$. Then

$$\rho_{W^u}(f^n(y), f^n(z)) \ge \nu \delta_1,$$

where $\nu$ is the maximum of the norm of $df$ over $M$.

By assumption, there are $C^1$ curves $\sigma_y$ and $\sigma_z$ which have length less than or equal to $r$, are tangent to $E^{cs}$ and connect $x$ to $y$ and $z$ respectively. By the partial hyperbolicity and center isometry,

$$\text{length}\,(f^n(\sigma_y)) \le c_2^{-1} \cdot \text{length}(\sigma_y), \quad \text{length}\,(f^n(\sigma_z)) \le c_2^{-1} \cdot \text{length}(\sigma_z)$$

and therefore,

$$\rho(f^n(y), f^n(z)) \le c_2^{-1}(\text{length}\,(\sigma_y) + \text{length}\,(\sigma_z)) \le 2c_2^{-1} r.$$

It follows that

$$c_1^{-1}\delta_0 \le \rho(f^n(y), f^n(z)) \le 2c_2^{-1} r$$

which leads to a contradiction for a sufficiently small $r$. □

The proof of the theorem uses this lemma and follows the line of the proof of the previous theorem. □

**5.4. Central foliation and normal hyperbolicity.** We describe a construction due to Hirsch, Pugh and Shub which allows one to treat central foliations in terms of normal hyperbolicity.

Let $f$ be a partially hyperbolic diffeomorphism in the narrow sense with a $df$-invariant splitting of the tangent bundle (2.7) satisfying (2.8). For $r > 0$ let $U_r(W^c(x)) \subset M$ be the tubular neighborhood of radius $r$ of the leaf $W^c(x)$. Consider the manifold that is the *disjoint* union

$$\mathcal{M}_r = \bigcup_{x \in M} U_r(W^c(x)). \tag{5.9}$$

For sufficiently small $\varepsilon$, $0 < \varepsilon < r$, the map $f$ induces a diffeomorphism $F : \mathcal{M}_\varepsilon \to \mathcal{M}_r$ which is normally hyperbolic to the submanifold

$$\mathcal{N} = \bigcup_{x \in M} W^c(x) \subset \mathcal{M}_\varepsilon.$$

The manifolds $\mathcal{M}_\varepsilon$ and $\mathcal{N}$ are not compact but complete. However, Theorems 5.2 and 5.3 extend to this situation (their proofs rely only on the existence of a tubular neighborhood of the normally hyperbolic manifold, uniform lower bound for the radius of injectivity of the exponential map, and uniform estimates (5.2); we have all this at our disposal since the manifold $M$ is compact and the central foliation $W^c$ is uniquely integrable). As a corollary of Theorem 5.2 we obtain the following statement.

**Theorem 5.9.** *Let $f\colon M \to M$ be a partially hyperbolic diffeomorphism in the narrow sense. Assume that the central distribution $E^c$ is integrable. Then $W^c(x) \in C^\ell$ for every $x \in M$ where the number $\ell$ is given by (5.8).*

**5.5. Robustness of the central foliation.** We describe another application of the Hirsch–Pugh–Shub construction and show that the central foliation $W^c$ survives under small perturbation of the system provided this foliation satisfies some specific requirements (for example, it is smooth).

Let $g$ be a $C^q$ diffeomorphism which is sufficiently close to $f$ in the $C^1$ topology.

**Theorem 5.10** (Hirsch, Pugh and Shub, [26], Theorem 7.5). *Assume that the central distribution $E^c$ for $f$ is integrable and the corresponding foliation $W^c$ is smooth. Then $g$ is partially hyperbolic in the narrow sense with integrable central distribution $E^c_g$.*

*Sketch of the proof.* The theorem is an immediate corollary of Theorems 5.12 and 5.14 below whose proofs exploit relatively complicated notions of plaque expansivity and pseudo orbits. To help us understand the role of these notions we present here an independent proof of Theorem 5.10.

If $d_{C^1}(f,g)$ is sufficiently small the diffeomorphism $g$ induces a diffeomorphism $G : \mathcal{M}_\varepsilon \to \mathcal{M}_r$, see (5.9). Theorem 5.3 applies to this situation and provides us with a submanifold $\mathcal{N}_G \subset \mathcal{M}_\varepsilon$, to which $G$ is normally hyperbolic. Moreover, given $r > 0$, if $d_{C^1}(f,g)$ is sufficiently small, then $\mathcal{N}_G$ lies in an $r$-neighborhood of $\mathcal{N} = \mathcal{N}_F$.

Consider the natural projection $\pi : \mathcal{M}_\varepsilon \to M$. Given $x \in M$, there exists a smooth immersed submanifold $W_g^c(x) \subset \pi(\mathcal{N}_G)$ through $x$ which is tangent to $E_g^c(y)$ at every point $y \in W_g^c(x)$. Moreover, these manifolds are invariant under $g$, i.e., $g(W_g^c(x)) = W_g^c(g(x))$. Thus, the result would follow if we could show that these manifolds form a partition of $M$, i.e., for any $x \in M$ and any $y \notin W_g^c(x)$ the intersection $W_g^c(x) \cap W_g^c(y)$ is empty.

To do this we shall use a result from [26] (see Theorem 6.1) which is an extension of Statement 3 of Theorem 5.3 to our situation.

**Lemma 5.11.** *There exists a homeomorphism $H\colon \mathcal{M}_\varepsilon \to \mathcal{M}_r$ which is $\delta$-close to the identity map in the $C^0$ topology and such that $H(\mathcal{N}_F) = \mathcal{N}_G$.* [4]

The desired result will now follow if we show that the map $h = \pi \circ H|\mathcal{N}_G \circ \tilde{\pi}$ is a homeomorphism, where $\tilde{\pi}$ lifts every point $x \in M$ to the point $x \in \mathcal{N}_F$. Note that $h$ is a *leaf conjugacy* , i.e., for any $x \in \mathcal{M}$ the points $h(f(x))$ and $g(h(x))$ lie on the same $W_g^c$-leaf. Moreover, $h$ is close to the identity map in the $C^0$ topology, i.e., $\rho(x, h(x)) \le \delta$ for every $x \in M$.

First, we shall show that $h$ is one-to-one. Assuming the contrary there exist $x, y \in M$ such that $h(x) = h(y) = z$. We have that $\rho(x,z) \le \delta$ and $\rho(y,z) \le \delta$ and hence, $\rho(x,y) \le 2\delta$. Since the foliation $W^c$ is smooth, there is a coordinate chart $U_x$ at $x$ of size $r > 0$ such that the local leaves of the central foliation can be viewed as parallel balls. Let $B^c(x)$ and $B^c(y)$ be those balls containing $x$ and $y$ respectively. We have that

$$d = \rho(B^c(x), B^c(y)) \le 2\delta. \tag{5.10}$$

There is a point $p \in W^u(x)$ such that the intersection $W^s(p) \cap B^c(y)$ is not empty and consists of a single point $q$ (see Figure 9). We have that either $\rho(x,p) \ge d/2$ or $\rho(p,q) \ge d/2$ and without loss of generality we may assume the former (otherwise one should apply the argument below to $f^{-1}$).

Observe that

$$\begin{gathered} \rho(f^n(p), f^n(q)) \le C\alpha^n, \quad \rho(f^n(q), f^n(y)) \le C\beta^n, \\ \rho(f^n(x), f^n(p)) \le C\gamma^n, \end{gathered} \tag{5.11}$$

where $C > 0$ and $0 < \alpha < \beta < \gamma$ are some constants and $\alpha < 1$, $\gamma > 1$. Note that $n$ can be taken arbitrarily large if the size of the perturbation

[4] The proof of this statement uses the assumption that the distribution $E_f^c$ is *uniquely* integrable.

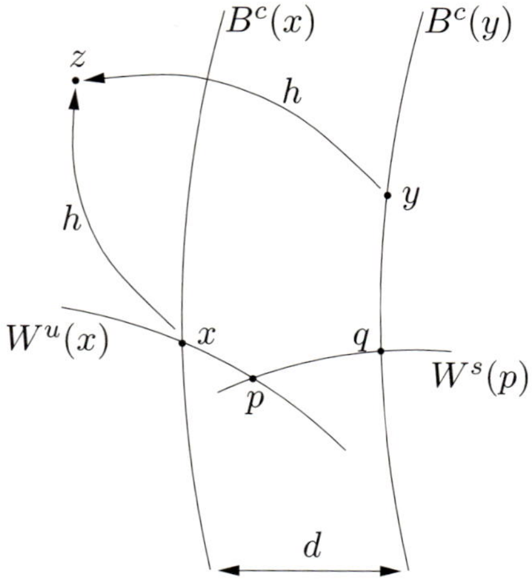


FIGURE 9. To the proof of Theorem 5.11.

(and hence, the number $d$) is sufficiently small. Consider the coordinate charts $U_{f^k(x)}$ at $f^k(x)$ of size $r$, $k = 0, 1, \dots, n$. By (5.11), we have that $f^k(y) \in U_{f^k(x)}$ for all $k$ and that

$$\rho(B^c(f^n(x)), B^c(f^n(y))) \geq C_1\gamma^n$$

for some constant $C_1 > 0$. On the other hand, the points $h(f^k(x))$ and $h(f^k(y))$ must lie on the central leaf of $g$ through $g^k(z)$ for all $k$ and hence, the distance $\rho(B^c(f^n(x)), B^c(f^n(y)))$ should not exceed $2\delta$. This is a contradiction.

We shall now show that $h$ is continuous. Otherwise there is a number $a > 0$ and a sequence of points $x_n$, converging to a point $x$, such that $\rho(h(x_n), h(x)) \geq a$ for all sufficiently large $n$. Since $h$ is $\delta$-close to the identity map (for some small $\delta > 0$), the number $a$ should be sufficiently small. By smoothness of the central foliation $W^c$ we obtain that

$$\rho(B^c(h(x_n)), B^c(h(x))) \geq a. \tag{5.12}$$

Fix $m > 0$ and consider the sequence of points $f^m(x_n)$ which converges to $f^m(x)$. No matter how big $m$ is one can always find $N > 0$ such that $\rho(f^m(x_n), f^m(x)) \leq \delta/2$ for all $n \geq N$. Therefore,

$$\rho(B^c(f^m(x_n)), B^c(f^m(x))) \leq \frac{\delta}{2}. \tag{5.13}$$

On the other hand, if $m$ is sufficiently big, by (5.12), we find that

$$\rho(B^c(g^m(h(x_n))), B^c(g^m(h(x)))) \geq 2\delta. \tag{5.14}$$

Since $h$ is a leaf conjugacy, $h(f^m(x_n)) \in W^c(g^m(h(x_n)))$ and $h(f^m(x)) \in W^c(g^m(h(x)))$. However, $h$ is $\delta$-close to the identity map and (5.13) contradicts (5.14).

Since $h$ is near the identity its surjectivity follows from its continuity. This completes the proof of the theorem. □

In general, we do not have the "luxury" of a smooth central foliation (so that (5.10) could be use). Moreover, even if the central foliation for $f$ were smooth the central foliation for a "typical" perturbation of $f$ would not be smooth. In [26], Hirsh, Pugh and Shub introduced a property of the central foliation for $f$ called *plaque expansivity*. This property is weaker than smoothness (see Theorem 5.12 below) but still guarantees integrability of the central distribution for sufficiently small perturbations of $f$. Unlike smoothness, it survives under small perturbations (see Theorem 5.14 below).

Let $W$ be a foliation of a compact smooth manifold $M$ whose leaves are $C^r$ smooth immersed submanifolds of dimension $k$. Given a point $x \in M$, we call the set $P(x) \subset W(x)$ a $C^r$ *plaque* of $W$ at $x$ if $P(x)$ is the image of a $C^r$ embedding of the unit ball $D \subset \mathbb{R}^k$ into $W(x)$. A *plaquation* $\mathcal{P}$ for $W$ is a collection of plaques such that every point $x \in M$ is contained in a plaque $P \in \mathcal{P}$.

Let $\{x_n\}_{n\in\mathbb{Z}}$ be a pseudo orbit for $f$ (see Section 4.8). We say that the pseudo orbit *respects* a plaquation $\mathcal{P}$ for $W$ if for every $n \in \mathbb{Z}$ the points $f(x_n)$ and $x_{n+1}$ lie in a common plaque $P \in \mathcal{P}$.

Assume that the foliation $W$ is invariant under a diffeomorphism $f$ of $M$. We say that $f$ is *plaque expansive* with respect to $W$ if there exists $\varepsilon > 0$ with the following property: if $\{x_n\}_{n\in\mathbb{Z}}$ and $\{y_n\}_{n\in\mathbb{Z}}$ are $\varepsilon$-pseudo orbits which respect $W$ and if $\rho(x_n, y_n) \leq \varepsilon$ for all $n \in \mathbb{Z}$ then $x_n$ and $y_n$ lie in a common plaque for all $n \in \mathbb{Z}$.

Note that plaque expansivity does not depend on the choice of either the Riemannian structure in $M$ or the plaquation $\mathcal{P}$ for $W$.

**Theorem 5.12** (Hirsch, Pugh and Shub, [26], Theorem 7.2). *Let $f$ be a partially hyperbolic diffeomorphism in the narrow sense. Assume that the central distribution $E^c$ is integrable and the central foliation $W^c$ is smooth. Then $W^c$ is plaque expansive.*

*Proof.* To illustrate the main idea of the proof let us assume, in addition, that the foliations $W^s$ and $W^u$ are smooth and jointly integrable. To establish plaque expansivity consider two $\varepsilon$-pseudo orbits $\{x_n\}_{n\in\mathbb{Z}}$ and $\{y_n\}_{n\in\mathbb{Z}}$, which respect $W^c$ and for which $\rho(x_n, y_n) \leq \varepsilon$ for all $n \in \mathbb{Z}$. Since the foliations $W^c$, $W^s$ and $W^u$ are smooth, for every $x \in M$ there is a local coordinate system near $x$ such that local leaves of these foliations are transversal parallel planes.

Assume now that the point $x_1$ does not lie in a local central manifold $V^c(y_1)$ and let $a > 0$ be the distance between the local leaves $V^c(x_1)$ and $V^c(y_1)$. Since $f(x_n)$ lies in the same plaque of $W^c$ as $x_{n+1}$ we have that $x_{n+1} \in f(V^c(x_n))$. Similarly, $y_{n+1} \in f(V^c(y_n))$. Applying the inverse map we obtain, in analogue, that $x_{n-1} \in f^{-1}(V^c(x_n))$ and $y_{n-1} \in f^{-1}(V^c(y_n))$. Since $f$ expands along $W^u$ and contracts along $W^s$ with an exponential rate it is easy to see that $\rho(V^c(x_n), V^c(y_n)) \geq \gamma^{|n|}a$ for some $\gamma > 1$ and either all $n > 0$ or all $n < 0$. However, this contradicts the fact that $\rho(x_n, y_n) \leq \varepsilon$ for all $n$.

In the general case when the foliations $W^s$ and $W^u$ are neither smooth nor jointly integrable we will exploit a similar idea and follow the proof in [26].

Passing to some power $f^n$, if necessary, we may assume that $\mu_1 < \frac{1}{2}$ and $\lambda_3 > 2$ (see (2.8)). Choose numbers $\gamma$ and $\gamma_0$ such that $\mu_1 < \gamma^{-1} < \frac{1}{2}$ and $\lambda_3 > \gamma > \gamma_0 > 2$.

For $p \in M$ and $v \in T_pM$ write $v = v^s + v^c + v^u$ with $v^s \in E^s(p)$, $v^c \in E^c(p)$ and $v^u \in E^u(p)$. Starting with a Lyapunov inner norm in $TM$ we shall use a new Finsler norm in $TM$ defined by $\|v\| = \max\{\|v^s\|, \|v^c\|, \|v^u\|\}$.

Given a smooth path $g\colon [0,1] \to M$ define its length by

$$L(g) = \int_0^1 \|g'(t)\|\, dt$$

and its $us$-length by

$$L_{us}(g) = \int_0^1 \max\{\|(g'(t))^u\|, \|(g'(t))^s\|\}\, dt.$$

Since the central distribution $E^c$ is uniquely integrable, any curve, which is everywhere tangent to the leaves of $W^c$, lies in a single leaf. This implies that $L_{us}(g) = 0$ if and only if $g$ maps into a leaf of $W^c$. For any $\nu > 0$ and $p, q \in M$ define

$$d^\nu_{us}(p, q) = \inf\{L_{us}(g)\colon L(g) \leq \nu\},$$

where the infinum is taken over all smooth paths $g$ from $p$ to $q$. We leave it to the reader to show that

$$\lim_{\nu,\nu' \to 0} \frac{d^\nu_{us}(p, q)}{d^{\nu'}_{us}(p', q')} = 1, \tag{5.15}$$

where we assume that $p' \in W^c(p)$, $q' \in W^c(q)$ and $\rho(p,p'), \rho(q,q') \le \min\{\nu, \nu'\}$. This is where smoothness of $W^c$ is essential. We need the following statement.

**Lemma 5.13.** *For any smooth path $g$,*

$$\max\{L_{us}(f^{-1} \circ g), L_{us}(f \circ g)\} \ge \frac{\gamma}{2} L_{us}(g). \tag{5.16}$$

*Proof.* Let

$$A(g) = \{t \in [0,1] \colon \|(g'(t))^u\| \ge \|(g'(t))^s\|\}, \quad B(g) = [0,1] - A(g).$$

Then

$$L_{us}(g) = \int_{A(g)} \|(g'(t))^u\|\, dt + \int_{B(g)} \|(g'(t))^s\|\, dt.$$

Since $df$ expands vectors in $E^u$ and contracts vectors in $E^s$ uniformly we have that

$$A(g) \subset A(f \circ g), \quad B(g) \subset B(f^{-1} \circ g).$$

Thereofre,

$$L_{us}(f \circ g) \ge \int_{A(f \circ g)} \|df(g'(t))^u(t)\|\, dt \ge \gamma \int_{A(g)} \|(g'(t))^u\|\, dt,$$

$$L_{us}(f^{-1} \circ g) \ge \int_{B(f^{-1} \circ g)} \|df^{-1}(g'(t))^s(t)\|\, dt \ge \gamma \int_{B(g)} \|(g'(t))^s\|\, dt.$$

Adding them together gives

$$L_{us}(f \circ g) + L_{us}(f^{-1} \circ g) \ge \gamma L_{us}(g).$$

This implies (5.16). □

We proceed with the proof of the theorem. Let $\{x_n\}_{n \in \mathbb{Z}}$ and $\{y_n\}_{n \in \mathbb{Z}}$ be two $\varepsilon$-pseudo orbits, which respect $W^c$ and for which $\rho(x_n, y_n) \le \varepsilon$ for all $n \in \mathbb{Z}$. It is easy to see that given $\nu > 0$ there exists $\nu' > 0$ such that $L(g) \le \nu$ for any path $g$ with $L(f \circ g) \le \nu'$ or $L(f^{-1} \circ g) \le \nu'$. It suffices to show that

$$\sigma = \sup_{n \in \mathbb{Z}} d^{\nu}_{us}(x_n, y_n) = 0.$$

Otherwise, given $\alpha > 0$ there is a number $m$ such that

$$1 \ge \frac{d^{\nu'}_{us}(x_m, y_m)}{\sigma} \ge 1 - \alpha. \tag{5.17}$$

Call $x = x_m$ and $y = y_m$ and observe that

$$\begin{aligned} d^{\nu'}_{us}(f(x), f(y)) &= \inf\{L_{us}(g) \colon g(0) = f(x), g(1) = f(y), L(g) \le \nu'\} \\ &\ge \inf\{L_{us}(f \circ g) \colon g(0) = x, g(1) = y, L(g) \le \nu\} \end{aligned}$$

and similarly for $d^{\nu'}_{us}(f^{-1}(x), f^{-1}(y))$. Hence, (5.16) yields

$$\max\{d^{\nu'}_{us}(f(x), f(y)),\ d^{\nu'}_{us}(f^{-1}(x), f^{-1}(y))\} \geq \frac{\gamma}{2} d^{\nu}_{us}(x,y). \tag{5.18}$$

Dividing (5.18) by $d^{\nu}_{us}(x,y)$ and using (5.15) we obtain for sufficiently small $\nu$,

$$\frac{d^{\nu}_{us}(x_{m+1}, y_{m+1})}{d^{\nu}_{us}(x_m, y_m)} \geq \frac{\gamma_0}{2} \quad \text{or} \quad \frac{d^{\nu}_{us}(x_{m-1}, y_{m-1})}{d^{\nu}_{us}(x_m, y_m)} \geq \frac{\gamma_0}{2}. \tag{5.19}$$

However, (5.19) is incompatible with (5.17) since $\gamma_0/2 > 1$ is fixed and $\alpha$ can be chosen arbitrary small. □

Let us consider the case when $df|E^c(x)$ acts as an isometry for every $x \in M$. By Theorem 5.7, the central distribution $E^c$ is integrable. One can show that in this case the central foliation $W^c$ is plaque expansive (see [26], Section 7).

**Theorem 5.14** (Hirsch, Pugh and Shub, [26], Theorem 7.1). *Let $f$ be a partially hyperbolic diffeomorphism in the narrow sense. If the central distribution $E^c$ for $f$ is integrable and $f$ is plaque expansive with respect to the central foliation $W^c$ then any diffeomorphism $g$ sufficiently close to $f$ is partially hyperbolic in the narrow sense, its central distribution $E^c_g$ is integrable and $g$ is plaque expansive with respect to its central foliation $W^c_g$.*

*Sketch of the proof.* As in the proof of Theorem 5.10 all we have to do is to show that the map $h$ defined there is a homeomorphism. Note that $h$ is indeed, a homeomorphism on each plaque of the central foliation.

First, we shall show that $h$ is one-to-one. Let $x, y \in M$ be such that $h(x) = h(y) = z$ but $y$ does not lie on the plaque of $W^c$ at $x$. Starting with the point $x_0 = x$ consider the points $h(f(x_0))$ and $g(h(x_0)) = g(z)$. Since $h$ is a leaf conjugacy and $g$ is close to $f$, these points lie on the same leaf of $W^c_g$ and, indeed, on the same plaque. Therefore, there is a unique point $x_1 = h^{-1} \circ g \circ h(x_0)$ which lies on the same plaque as $f(x_0)$. Continuing in a similar fashion one can construct a sequence of points $\{x_n\}_{n\in\mathbb{Z}}$ which is a pseudo orbit for $f$ and respects the plaquation for $W^c$. Starting now with the point $y_0 = y$, one can construct another pseudo orbit for $f$ which respects the plaquation for $W^c$. Clearly, the distance $\rho(x_n, y_n)$ is of order $\varepsilon$ for all $n$. By plaque expansivity of $f$ we conclude that the points $x$ and $y$ must lie in a common plaque for $W^c$.

We shall now show that $h$ is continuous. Otherwise there is a number $a > 0$ and a sequence of points $x_m$, converging to a point $x$, such that

$$\rho(h(x_m), h(x)) \geq a \tag{5.20}$$

for all sufficiently large $m$. Fix $m$ and starting with $x_m$ consider the pseudo orbit $\{x_{m,n}\}_{n\in\mathbb{Z}}$ for $f$ constructed above. By the diagonal process one can choose a subsequence $x_{m_k,n}$ which converges, for each $n \in \mathbb{Z}$, to a point $x_n$ as $k \to \infty$. Clearly, $x_n$ is a pseudo orbit for $f$ through $x$. Hence, given $\varepsilon > 0$ and $L > 0$ we have

$$\rho(x_n, g^n(h(x))) \le \rho(x_n, x_{m_k,n}) + \rho(x_{m_k,n}, g^n(h(x))) \le \varepsilon + \varepsilon = 2\varepsilon$$

if $k$ is sufficiently large and $-L \le n \le L$. However, this contradicts (5.20) due to partial hyperbolicity (see [26], Theorem 7.1 for more details).

Surjectivity of $f$ is a corollary of its continuity since $h$ is near identity.

In order to prove plaque expansivity of $g$ consider two pseudo orbits $\{x_n\}_{n\in\mathbb{Z}}$ and $\{y_n\}_{n\in\mathbb{Z}}$ for $g$ which respect the central foliation $W_g^c$ and such that $\rho(x_n, y_n) \le \frac{1}{2}\varepsilon$. The sequences $\{x_n' = h^{-1}(x_n)\}_{n\in\mathbb{Z}}$ and $\{y_n' = h^{-1}(y_n)\}_{n\in\mathbb{Z}}$ are pseudo orbits for $f$ which respect the central foliation $W^c$ and such that $\rho(x_n', y_n') \le \varepsilon$ (provided that $g$ is sufficiently close to $f$). By plaque expansivity of $f$ we conclude that $x_n'$ and $y_n'$ lie in a common plaque for $W^c$ for all $n$ and hence, $x_n$ and $y_n$ lie in a common plaque for $W_g^c$ for all $n$. □

**5.6. Weak integrability of the central foliation.** As we mentioned above the central distribution may not be integrable. However, it is often integrable in some weak sense and this weak integrability persists under small perturbations. Following [10] we call a continuous distribution $E$ on a compact smooth manifold $M$ *weakly integrable* if for each point $x \in M$ there is an immersed complete $C^1$ manifold $W(x)$ which contains $x$ and is everywhere tangent to $E$, i.e., $T_yW(y) = E(y)$ for each $y \in W(x)$. We call $W(x)$ an integral manifold of $E$ through $x$. Note that *a priori* the integral manifolds $W(x)$ may be self-intersecting and may not form a partition of $M$.

**Theorem 5.15** (Brin, Burago and Ivanov, [10]). *Let $f$ be a partially hyperbolic in the narrow sense diffeomorphism of $M$. Assume that the distributions $E_f^{cs}$, $E_f^{cu}$ and $E_f^c$ are weakly integrable. Then there is a $C^1$ neighborhood $\mathcal{U}$ of $f$ such that every $g \in \mathcal{U}$ is a partially hyperbolic in the narrow sense diffeomorphism whose distributions $E_g^{cs}$, $E_g^{cu}$ and $E_g^c$ are weakly integrable.*

*Proof.* We will prove the weak integrability of $E_g^{cu}$. The weak integrability of $E_g^{cs}$ follows by reversing the time. The integral manifolds of $E_g^c$ can be constructed by intersecting the integral manifolds of $E_g^{cu}$ with the integral manifolds of $E_g^{cs}$.

Fix $x \in M$. For each $n > 0$ let $W_n = g^n(W_f^{cu}(g^{-n}(x)))$ where $W_f^{cu}(y)$ is the integral manifold of $E_f^{cu}$ through $y \in M$. It follows from Theorem 3.4 that the distributions $E_g^s$, $E_g^u$ and $E_g^c$ depend continuously on $g$ in the $C^1$

topology and that $E_f^{cu}(y)$ is uniformly close to $E_g^{cu}(y)$ and hence, is uniformly transverse to $E_g^s(y)$. Moreover, the hyperbolicity constants of $f$ (see (2.6)) are $\delta$-close to those of $g$ for a given sufficiently small $\delta > 0$.

It follows from (2.7) that for any $\alpha > 0$ there is $C(\alpha) > 0$ such that for any $N > 0$ and any tangent vector $v \in T_xM$ with the distance $d(\frac{v}{\|v\|}, E_f^s(x)) \geq \alpha$ we have

$$\|d(df^n v, E_f^{cu}(f^n(x)))\| \leq C(\alpha)\Big(\frac{\mu_1}{\lambda_2}\Big)^n \|v\|.$$

This implies that the tangent space to $W_n$ is exponentially close to $E_g^{cu}$, i.e.,

$$T_{g^n(y)}(g^n(W_f^{cu}(g^{-n}(x)))) \to E_g^{cu}(g^n(y))$$

uniformly in $y \in W_f^{cu}(g^{-n}(x))$ and exponentially in $n$ with rate $(\mu_1+\delta)/(\lambda_2-\delta)$. The sequence $W_n$ is precompact in the compact-open $C^1$ topology and therefore, has a convergent subsequence. The limit is an integral manifold of $E_g^{cu}$ passing through $x$. □

**Theorem 5.16** (Brin, Burago and Ivanov, [10]). *Let $\{f_n\}_{n\geq 0}$ be a sequence of partially hyperbolic in the narrow sense diffeomorphisms of $M$. Assume that*

(1) *$f_n \to g$ in the $C^1$ topology;*
(2) *the distributions $E_{f_n}^{cs}$, $E_{f_n}^{cu}$ and $E_{f_n}^c$ are weakly integrable for all $n$;*
(3) *all $f_n$ have the same hyperbolicity constants* (2.6).

*Then $g$ is partially hyperbolic in the narrow sense and the distributions $E_g^{cs}$, $E_g^{cu}$ and $E_g^c$ are weakly integrable.*

*Proof.* Fix $x \in M$ and let $v \in E_{f_n}^s(x)$ be a sequence of unit vectors converging to a vector $v \in T_xM$. Then $\lambda_1^n \leq \|df^n v\| \leq \mu_1^n$ and hence, $v \in E_g^s(x)$. Similar arguments apply to $E^c$ and $E^u$. It follows that

$$\lim_{n\to\infty} E_{f_n}^\alpha(x) = E_g^\alpha(x)$$

for $\alpha = s, c, u$. Hence, $g$ is partially hyperbolic in the narrow sense. We will prove weak integrability of $E_g^{cu}$. The integrability of $E_g^{cs}$ and $E_g^c$ follows by reversing the time and taking intersections. Let $W_n$ be an integral manifold of $E_{f_n}^{cu}$ passing through $x \in M$. Since the distribution $E_g^{cu}$ depends continuously on $g$ in the $C^1$ topology, the sequence $W_n$ is precompact in the compact-open $C^1$ topology and therefore, has a convergent subsequence. The limit is an integral manifold of $E_g^{cu}$ passing through $x$. □

## 6. Intermediate Foliations

Consider a diffeomorphism $f$ of class $C^q$ of a compact Riemannian manifold $M$ admitting a $df$-invariant splitting (4.41) satisfying (4.42) and (4.43).

Fix a number $k$, $1 < k < t$ with $\mu_k < 1$. In this section we will discuss the integrability problem for the invariant distribution $E_k$ called the *intermediate* distribution.

**6.1. Non-integrability of intermediate distributions.** In general, one should not expect $E_k$ to be integrable as the following example illustrates.

**Example 6.1.** [Smale, [48], see also [29], Section 17.3]

Consider the *Heisenberg group* of matrices

$$H = \left\{ \begin{pmatrix} 1 & x & z \\ 0 & 1 & y \\ 0 & 0 & 1 \end{pmatrix} : (x, y, z) \in \mathbb{R} \right\}$$

with the usual matrix multiplication: in $(x, y, z)$ coordinates it is given by

$$(x_1, y_1, z_1) \times (x_2, y_2, z_2) = (x_1 + x_2, y_1 + y_2, z_1 + z_2 + x_1 y_2).$$

The center of $H$ is the one-parameter subgroup

$$\begin{pmatrix} 1 & 0 & z \\ 0 & 1 & 0 \\ 0 & 0 & 1 \end{pmatrix}.$$

Thus, $H$ is a three-dimensional, simply connected, nonabelian and nilpotent group. The Lie algebra of $H$ is given by

$$\mathfrak{L}(H) = \left\{ \begin{pmatrix} 0 & x & z \\ 0 & 0 & y \\ 0 & 0 & 0 \end{pmatrix} : (x, y, z) \in \mathbb{R} \right\}$$

with generators

$$X = \begin{pmatrix} 0 & 1 & 0 \\ 0 & 0 & 0 \\ 0 & 0 & 0 \end{pmatrix}, \quad Y = \begin{pmatrix} 0 & 0 & 0 \\ 0 & 0 & 1 \\ 0 & 0 & 0 \end{pmatrix}, \quad Z = \begin{pmatrix} 0 & 0 & 1 \\ 0 & 0 & 0 \\ 0 & 0 & 0 \end{pmatrix}.$$

We have the Lie bracket $[X, Y] = Z$ while all other brackets of generators are zero.

Let $G = H \times H$ be the Lie group with generators $X_1, Y_1, Z_1, X_2, Y_2, Z_2$ such that $[X_i, Y_i] = Z_i$ and all other brackets of generators are zero. The Lie algebra of $G$ is

$$\mathfrak{L}(G) = \left\{ \begin{pmatrix} A & 0 \\ 0 & B \end{pmatrix} : A, B \in \mathfrak{L}(H) \right\}.$$

The group $H$ has an obvious integer lattice of matrices with entries in $\mathbb{Z}$ which generates an integer lattice in $G$ (recall that a lattice $\Gamma$ is a discrete subgroup of $G$ for which the factor $\Gamma\backslash G$ is compact; see Section 2.1). We need another lattice in $G$ however.

Consider the number field $\mathbb{K} = \{a + b\sqrt{5}\colon a, b \text{ are rational numbers}\}$. It possesses a unique nontrivial automorphism $\sigma$ such that $\sigma(a + b\sqrt{5}) = a - b\sqrt{5}$.

Let $\Gamma$ be the subgroup of $G$ given by $\exp_{\mathrm{Id}} \gamma$, where $\exp_{\mathrm{Id}}\colon \mathfrak{L}(G) \to G$ is the exponential map and $\gamma \subset \mathfrak{L}(G)$ is given by

$$\gamma = \left\{ \begin{pmatrix} A & 0 \\ 0 & \sigma(A) \end{pmatrix} : A \in \mathfrak{L}(H) \text{ with entries in the algebraic integers in } \mathbb{K} \right\}$$

with $\sigma(A)_{ij} = \sigma(A_{ij})$. It can be shown that $\Gamma$ is a lattice (see [29], Section 17.3). Define a Lie algebra automorphism $\Phi$ on $\mathfrak{L}(G)$ by

$$\Phi(X_1) = \lambda_1 X_1, \quad \Phi(Y_1) = \lambda_1^2 Y_1, \quad \Phi(Z_1) = \lambda_1^3 Z_1,$$

$$\Phi(X_2) = \lambda_1^{-1} X_2, \quad \Phi(Y_2) = \lambda_1^{-2} Y_2, \quad \Phi(Z_2) = \lambda_1^{-3} Z_2,$$

where $\lambda_1 = \frac{3+\sqrt{5}}{2}$ and $\lambda_2 = \frac{3-\sqrt{5}}{2}$. There exists a unique automorphism $F\colon G \to G$ with $dF|\,\mathrm{Id} = \Phi$. Since $\lambda_1$ and $\lambda_2$ are units in $\mathbb{K}$, that is integers whose inverses are also integers, and $\sigma(\lambda_1) = \lambda_2$ we have that $F(\Gamma) = \Gamma$. Thus, $F$ projects to an Anosov diffeomorphism $f$ of $\Gamma\backslash G$.

The invariant splitting for $f$ is $T(\Gamma\backslash G) = E^s \oplus E^u$, where $E^s$ is the three-dimensional distribution generated by $X_2$, $Y_2$ and $Z_2$ and $E^u$ is the three-dimensional distribution generated by $X_1$, $Y_1$ and $Z_1$. Observe that $E^u = P \oplus Q$ where $P$ is the two-dimensional distribution generated by $X_1, Y_1$ and $Q$ is the one-dimensional distribution generated by $Z_1$. The distribution $P$ is intermediate and is not integrable. To see this note that the generators $X_1$, $Y_1$ and $Z_1$ induce three vector fields $x_1$, $y_1$ and $z_1$ on $g \in \Gamma\backslash G$ such that $x_1(g), y_1(g) \in P(g)$ and $z_1(g) \in Q(g)$ for any $g \in \Gamma\backslash G$. Since the distribution $P$ is smooth, by the Frobenius theorem, its integrability would imply that the Lie bracket $[x_1, y_1]$ of vector fields $x_1$ and $y_1$ lies in $P$. However, this is not true since $[X_1, Y_1] = Z_1$.

It follows from Theorem 5.16 that non-integrability in the above example is an open property.

**6.2. Invariant families of local manifolds.** One can always view the invariant intermediate distribution $E_k$, $1 < k < t$, as a central distribution in the following splitting of the tangent bundle

$$TM = (\oplus_{j=1}^{k-1} E_j) \oplus E_k (\oplus_{j=k+1}^{t} E_j).$$

and apply results of Sections 5.4 and 5.5. First, we establish the class of smoothness of the leaves of the foliation $W_k$ (provided the distribution $E_k$ is integrable). Denote by $\ell_k^1$ and $\ell_k^2$ the biggest number such that

$$\mu_{k-1} < \lambda_k^{\ell_k^1}, \quad \mu_k < \lambda_{k+1}^{\ell_k^2}, \quad \ell = \min\{\ell_k^1, \ell_k^2\}$$

(recall that the numbers $\lambda_k$ and $\mu_k$ satisfy (4.42) and (4.43)). Applying Theorem 5.9 we obtain the following result.

**Theorem 6.2.** *Assume that the distribution $E_k$ is integrable. Then the leaves of the corresponding intermediate invariant foliation $W_k$ are of class of smoothness $C^\ell$.*

Let $g$ be a $C^q$ diffeomorphism which is sufficiently close to $f$ in the $C^1$ topology. By Theorem 3.4, $g$ possesses an invariant distribution $(E_k)_g$ corresponding to $E_k$. Applying Theorem 5.14 requires an additional assumption that $\lambda_{k+1} > 1$. Of course, this means that $f$ is an Anosov diffeomorphism.

**Theorem 6.3.** *Assume that the foliation $W_k$ is plaque expansive (in particular, smooth). Assume also that $\lambda_{k+1} > 1$. Then the distribution $(E_k)_g$ is integrable and the corresponding foliation $(W_k)_g$ is plaque expansive.*

Since the diffeomorphism $f$ in the last theorem is Anosov, so is $g$. By the structural stability theorem, $f$ and $g$ are topologically conjugate by a Hölder homeomorphism $h$ which is close to the identity map. It follows that $h(W_k)$ is a $g$-invariant foliation whose leaves are Hölder continuous submanifolds. Theorem 6.3 claims that the leaves of this foliation are indeed smooth of class $C^\ell$.

Even if the distribution $E_k$ is integrable its leaves may not be of class $C^q$ but less. To explain this phenomenon consider the linear map of the plane $A(x, y) = (\lambda x, \mu y)$ where $0 < \lambda < \mu < 1$ are some numbers. The origin is the attracting fixed point. The $x$-axis can be geometrically characterized as consisting of points $P$ for which

$$\rho(0, A^n P) \le \lambda^n \rho(0, P).$$

On the other hand any curve

$$\gamma_C = \{(x, y) : x = C y^{\frac{\log \mu}{\log \lambda}}\}$$

is invariant under $A$ and consists of points $P$ for which

$$\rho(0, A^n P) \le \mu^n \rho(0, P).$$

Note that these curves are only finitely differentiable except for the $y$-axis (corresponding to $C = 0$). Therefore, there is no "obvious" choice of a local leaf and "little" chances are that the intermediate foliation will "pick" the leaf that is infinitely differentiable. Surprisingly, if $\mu_k < 1$ and some special *non-resonance condition* holds, this simple picture is realizable: the distribution $E_k$ admits an invariant family of local manifolds $\{V_k(x)\}_{x\in M}$ which are as smooth as the map $f$ is but may not comprise a foliation.

**Theorem 6.4** (Pesin, [37]). *Fix $k$ such that $0 < \lambda_k \le \mu_k < 1$. Assume that the following non-resonance condition holds:*

$$q \ge N = [\log \lambda_1 / \log \mu_k] + 1 \text{ and for every } j = 1, \dots, N$$
$$[(\lambda_k)^j, (\mu_k)^j] \cap [\lambda_i, \mu_i] = \emptyset, \quad 1 \le i < k. \tag{6.1}$$

*Then for every $x \in M$ there exists a local submanifold $V_k(x)$ satisfying:*

(1) *$x \in V_k(x)$ and $T_x V_k(x) = E_k(x)$;*
(2) *$f(V_k(x)) \subset V_k(f(x))$;*
(3) *$V_k(x) \in C^q$;*
(4) *for any $x \in M$ the collection of local manifolds $\{V_k(x)\}_{x\in M}$ is the only collection of $C^N$ local manifolds that satisfies $T_x V_k(x) = E_k(x)$, $f(V_k(x)) \subset V_k(f(x))$, and*

$$\sup_{1\le s\le N} \sup_{x\in M} \|d^s V_k(x)\| \le \mathit{const.}$$

**Remark 6.5.** The non-integrable intermediate distribution $P$ for the diffeomorphism $f$ in Example 6.1 does not satisfy the non-resonance condition.

*Proof of the theorem.* We will find the local manifold $V_k(x)$ in the following form

$$V_k(x) = \exp_x\{(v, \psi_k(x, v)) \colon v \in B_k(r_0)\},$$

where $\psi_k(x, \cdot) \colon B_k(r_0) \to E(x)$ is a smooth map which satisfies $\psi_k(x, 0) = 0$ and $d\psi_k(x, 0) = 0$. Here $B_k(r_0)$ is the ball in $E_k(x)$ of radius $r_0$ centered at the origin and

$$E(x) = \bigoplus_{i=1, i\ne k}^{t} E_i(x).$$

The number $r_0$ is sufficiently small and is independent of $x$. To construct the function $\psi_k(x, \cdot)$ we proceed as in the proof of Theorem 4.1 and consider the map

$$\widetilde{f}_x = \exp_{f(x)}^{-1} \circ f \circ \exp_x \colon B_k(r_0) \times B(r_0) \to T_{f(x)} M,$$

which is well-defined if $r_0$ is sufficiently small. Here $B(r_0)$ is the ball in $E(x)$ of radius $r_0$ centered at the origin. We assume that a Lyapunov inner metric is introduced (see Section 2.2) so that we can write the map $\widetilde{f}_x$ in the form (4.7) with $A(x) = df|E_k(x)$ and $B(x) = df|E(x)$ satisfying

$$\|A(x)\| \le \mu_k, \quad (\|B(x)^{-1}\|)^{-1} \ge \gamma \tag{6.2}$$

and with

$$0 < \mu_k < 1, \quad \mu_k^N/\gamma < 1 \tag{6.3}$$

for some $\gamma > 0$. Furthermore,

$$g_x \colon B_k(r_0) \times B(r_0) \to E_k(f(x)), \quad h_x \colon B_k(r_0) \times B(r_0) \to E(f(x))$$

satisfy (4.9)–(4.11).

Since the local manifold $V_k(x)$ is invariant we have $f(V_k(x)) \subset V_k(f(x))$. Applying (4.7) we obtain the following equation for the function $\psi_k(x,\cdot)$:

$$\begin{aligned} &B(x)\psi_k(x,v) + h_x(v, \psi_k(x,v)) \\ &\qquad = \psi_k(f(x), A(x)v + g_x(v, \psi_k(x,v))), \quad v \in B_k(r_0). \end{aligned} \tag{6.4}$$

As the function $\psi_k(x,v)$ is of class $C^q$ with $q \geq N$ one can write

$$\begin{aligned} \psi_k(x,v) =& \psi_k(x,0) + d\psi_k(x,0)v + \cdots \\ &+ d^N\psi_k(x,0)(v,\dots,v) + \widetilde{\psi}_k(x,v) \\ =& t_k(x,v) + \widetilde{\psi}_k(x,v). \end{aligned} \tag{6.5}$$

Here $d^s\psi_k(x,0)$, $s = 0\dots N$, are symmetric $s$-forms on the manifold $M$. Substituting (6.5) into (6.4) and equating coefficients of "equal powers of $v$" (i.e., the corresponding $s$-forms) we come to the following system of equations on the manifold $M$

$$\begin{aligned} B_i(x)d^s\psi_i(x,0) =& d^s\psi_i(f(x),0)A^s(x) \\ &+ T_i(x, d^\ell\psi_k(x,0), d^j g_x(0,0), d^j h_x(0,0)), \end{aligned}$$

where $s = 0,\dots,N$, $i = 1,\dots,t$, $i \neq k$, $0 \leq \ell$, $j \leq s-1$, and $B_i(x) = df|E_i(x)$. We shall solve this equation by induction on $s$. Consider the Banach space of continuous symmetric $s$-forms on $M$ and for each $i$ define the linear operator $\Omega_i$ by

$$(\Omega_i\omega)(x) = (B_i(x))^{-1}\omega(f(x))A^s(x).$$

The non-resonance condition implies that either the operator $\Omega_i$ or its inverse is of norm (induced by the norm in the space of symmetric $s$-forms) strictly less than one. Therefore, the operator $(I-\Omega_i)^{-1}$ (or $(I-\Omega_i^{-1})^{-1}$) is well defined and bounded. The functions $T_i$ depend only on $d^\ell\psi_i(x,0)$ with $\ell \leq s-1$ and these $(s-1)$-forms are already defined. Therefore,

$$d^s\psi_i(x,0) = (I-\Omega_i)^{-1}T_i \quad (\text{or } d^s\psi_i(x,0) = (I-\Omega_i^{-1})^{-1}T_i).$$

Finally, for $s=0$ we have $\psi(x,0) = (x,0)$.

Choose the function $\widetilde{\psi}_k(x,v)$, $v \in B_k(r_0)$, such that $\widetilde{\psi}_k(x,0) = (x,0)$ for every $x \in M$, $\widetilde{\psi}_k(x,v)$ is $q$ times differentiable, and $d^i\widetilde{\psi}_k(x,0) = (x,0)$

for $i = 1, \dots, N$. It follows from (6.4) and (6.5) that the function $\widetilde{\psi}_k(x,v)$ satisfies the following equation

$$B(x)\widetilde{\psi}_k(x,v) + \widetilde{h}_x(v,\psi_k(x,v)) = \widetilde{\psi}_k(f(x), A(x)v + \widetilde{g}_x(v,\psi_k(x,v))), \tag{6.6}$$

where

$$\widetilde{g}_x\colon B_k(r_0)\times B(r_0)\to E_k(f(x)),\quad \widetilde{h}_x\colon B_k(r_0)\times B(r_0)\to E(f(x))$$

are such that

$$\widetilde{g}_x(0,0)=0,\quad \widetilde{h}_x(0,0)=0,\quad d\widetilde{g}_x(0,0)=0,\quad d^i\widetilde{h}_x(0,0)=0, \tag{6.7}$$

for $i = 1, \dots, N$ and

$$\|d\widetilde{g}_x(v,w)\|\le C_1,\quad \|d^{N+1}\widetilde{h}_x(v,w)\|\le C_1, \tag{6.8}$$

where $(v,w)\in B_k(r_0)\times B(r_0)$ and $C_1>0$ is a constant. The desired result is now an immediate consequence of the following lemma.

**Lemma 6.6.** *Under the conditions* (6.2), (6.3), (6.7) *and* (6.8), *Equation* (6.6) *has a unique solution* $\widetilde{\psi}_k(x,v)$ *which is a continuous function in* $x$ *on* $M$, *is* $q$*-times continuously differentiable in* $v$ *and*

$$\widetilde{\psi}_k(x,0)=0,\quad d^i\widetilde{\psi}_k(x,0)=0$$

*for* $i = 1, \dots, N$.

*Proof of the lemma.* Using (4.13) and (4.14) one can rewrite Equation (6.6) in the following form

$$B_m\widetilde{\psi}_{k,m}(v)+\widetilde{h}_m(v,\psi_{k,m}(v))=\widetilde{\psi}_{k,m+1}(A_mv+\widetilde{g}_m(v,\psi_{k,m}(v))),\quad v\in B_k(r_0),$$

where by (6.2),

$$\|A_m\|\le\mu_k,\quad (\|B_m{}^{-1}\|)^{-1}\ge\gamma, \tag{6.9}$$

and by (6.7) and (6.8) for $i = 1, \dots, N$ and $(v,w)\in B_k(r_0)\times B(r_0)$,

$$\begin{gathered}\widetilde{g}_m(0,0)=0,\quad \widetilde{h}_m(0,0)=0,\quad d\widetilde{g}_m(0,0)=0,\quad d^i\widetilde{h}_m(0,0)=0,\\ \|d\widetilde{g}_m(v,w)\|\le C_2,\quad \|d^{N+1}\widetilde{h}_m(v,w)\|\le C_2,\end{gathered} \tag{6.10}$$

where $C_2>0$ is a constant.

The proof of the desired result is now a modification of the argument in the proof of Theorem 4.3. Indeed, consider the space $\Gamma$ with the norm $\|\cdot\|_\kappa$

and the operator $\Phi(y,z)$. Using (6.9), (6.10), and (6.3) we find that

$$\begin{aligned}
\|\Phi(y,z)\|_\kappa &= \sup_{m\geq 0}(\kappa^{-m}\|\Phi(y,z)(m)\|) \leq \sup_{m\geq 0}(\kappa^{-m}\|z(m)\|) \\
&+ \sup_{m\geq 0}\left\{\kappa^{-m}\left[\left(\prod_{i=0}^{m-1}\|A_i\|\right)\|y\| + \sum_{n=0}^{m-1}\left(\prod_{i=n+1}^{m-1}\|A_i\|\right)C_2\|z(n)\|\right.\right. \\
&+ \left.\left.\sum_{n=0}^{\infty}\left(\prod_{i=0}^{n}\|B_{i+m}^{-1}\|\right)C_2\|z(n+m)\|^N\right)\right]\Bigg\} \\
&\leq \|z\|_\kappa + \sup_{m\geq 0}(\kappa^{-m}\mu_k{}^m)\|y\| \\
&+ \sup_{m\geq 0}\left(\kappa^{-m}C_2\|z\|_\kappa\left[\sum_{n=0}^{m-1}(\mu_k)^{m-n-1}\kappa^n + \sum_{n=0}^{\infty}\gamma^{-(n+1)}\kappa^{N(m+n)}\right]\right) \\
&\leq \|y\| + C_3\|z\|_\kappa,
\end{aligned}$$

where $C_3 > 0$ is a constant. Therefore, the map $\Phi$ is well-defined and $\Phi(0,0) = (0,0)$. One can now show that the map $\Phi$ is continuously $q$-times differentiable and $\partial_x\Phi(y,0) = -\mathrm{Id}$. The lemma follows from the Implicit Function Theorem. We leave the details to the reader. □

This completes the proof of the theorem. □

**6.3. Insufficient smoothness of intermediate foliations.** The following example illustrates the possible lack of smoothness of leaves for intermediate distributions. Consider an automorphism $A$ of the torus $\mathbb{T}^3$ with eigenvalues $\lambda_i$, $i = 1,2,3$, such that $0 < \lambda_1 < \lambda_2 < 1 < \lambda_3$. We have an invariant splitting

$$T\mathbb{T}^3 = \bigoplus_{i=1}^{3} E_{i,A}.$$

Assume that the number $\log\lambda_1/\log\lambda_2$ is not an integer, i.e., the above non-resonant condition holds. Let $N = [\log\lambda_1/\log\lambda_2] + 1$.

Consider the foliation $W_{2,A}$ associated to $E_{2,A}$. Any $C^\infty$ diffeomorphism $f$, which is sufficiently $C^1$-close to $A$, possesses an invariant foliation $W_{2,f}$ which is tangent to $E_{2,f}$ and whose leaves are of class $C^{N-1}$, see Theorem 6.2. We will show that in general, the leaves $W_{2,f}(x)$ cannot be more than $C^{N-1}$ smooth for a "large" set of points $x \in M$. Hence, they are different from the local submanifolds given by the preceding theorem, since these submanifolds are of class $C^N$ (indeed, of class $C^\infty$ in this particular case).

**Theorem 6.7** (Jiang, de la Llave and Pesin, [28]). *In any neighborhood $\eta$ of $A$ in the space* $\mathrm{Diff}^1(\mathbb{T}^3)$ *there exists $G \in \eta$ such that*

(1) *$G$ is a $C^\infty$ diffeomorphism and topologically conjugate to $A$;*
(2) *$G$ admits an invariant splitting*

$$T\mathbb{T}^3 = \bigoplus_{i=1}^{3} E_{i,G}$$

*with $E_{i,G}$ close to $E_{i,A}$ and integrable; the integral manifold $W_{i,G}(x)$ passing through $x$ is of class $C^{N-1}$ but not $C^N$ for some $x \in \mathbb{T}^3$;*
(3) *the set of points $\{x\colon W_{i,G}(x)$ is not of class $C^N\}$ is a residual subset of $\mathbb{T}^3$.*

*Proof.* We define the perturbation $G$ in the form $G = \pi \circ (\widetilde{A} + \widetilde{h})$ where $\pi\colon \mathbb{R}^3 \to \mathbb{T}^3$ is the usual projection, $\widetilde{A}$ is the lift of $A$, and $\widetilde{h}\colon \mathbb{R}^3 \to \mathbb{R}^3$ is a map. We describe $\widetilde{h}$ in the standard $\{e_1, e_2, e_3\}$-coordinate system where $e_i$ are eigenvectors of $A$.

Fix a point $p_0 = (0, \beta_0, 0) \in \mathbb{R}^3$ with sufficiently small $\beta_0 > 0$ and a small number $\varepsilon > 0$ such that the sets

$$\pi(B(0,\varepsilon)),\ A\pi(B(p_0,\varepsilon)),\ \pi(B(p_0,\varepsilon)),\ A^{-1}\pi(B(p_0,\varepsilon)),\dots,A^{-K}\pi(B(p_0,\varepsilon))$$

are all disjoint for some $K > 0$. For sufficiently small $\varepsilon$ and $\beta_0$ the number $K$ can be chosen so large that

$$\left(\frac{\lambda_2}{\lambda_1}\right)^K \frac{1}{8\lambda_1} > L. \tag{6.11}$$

We also require that $\widetilde{A}(B(p_0,\varepsilon))$ and $B(p_0,\varepsilon)$ be contained in the same fundamental domain $D$. Define a smooth function $f\colon D \to \mathbb{R}$ such that

1. $|f(\alpha,\beta,\gamma)| \le \delta$ and $\|df(\alpha,\beta,\gamma)\| \le \delta$ where $\delta$ is a constant satisfying
$$0 < \delta < \max\left\{\frac{\lambda_1\varepsilon}{4},\ \frac{1}{4}(\lambda_2 - \lambda_1)\right\};$$
2. $|\frac{\partial f}{\partial \beta}(\alpha,\beta_0,0)| \ge \frac{\delta}{4}$ for $|\alpha| \le \frac{\varepsilon}{3}$;
3. $f(\alpha,\beta,\gamma) = 0$ for $(\alpha,\beta,\gamma) \notin B(p_0,\varepsilon)$.

For all $(\alpha,\beta,\gamma) \in \mathbb{R}^3$ set

$$\widetilde{h}(\alpha,\beta,\gamma) = (f(\alpha,\beta,\gamma),0,0) \pmod 1$$

and then

$$\widetilde{G} = \widetilde{A} + \widetilde{h}, \quad G = \pi \circ (\widetilde{A} + \widetilde{h}).$$

If $\delta$ is sufficiently small then $G$ is a $C^\infty$ Anosov diffeomorphism of $\mathbb{T}^3$ which is topologically conjugate to $A$. The first statement follows. $G$ possesses a $dG$-invariant splitting $T\mathbb{T}^3 = \bigoplus_{i=1}^3 E_{i,G}$ and Mather's spectrum of $G$ satisfies the non-resonance condition. We denote the corresponding $d\widetilde{G}$-invariant splitting by $T\mathbb{R}^3 = \bigoplus_{i=1}^3 E_{i,\widetilde{G}}$ so that $d\pi E_{i,\widetilde{G}} = E_{i,G}$, $i = 1, 2, 3$.

It is easy to see that

$$G(x) = A(x)$$

for $x \notin \pi(B(p_0, \varepsilon))$. Since the origin is a fixed point of $\widetilde{A}$ and $\widetilde{G} = \widetilde{A}$ near the origin we conclude that

$$E_{2,\widetilde{G}}(0) = E_{2,\widetilde{A}}(0) = \{t\, e_2\colon\ t \in \mathbb{R}\}$$

and that the local smooth manifold at 0 is

$$V_{2,\widetilde{G}}(0) = V_{2,\widetilde{A}}(0) = \{t\, e_2\colon\ t \in (a, -a) \text{ for some } a > 0\}. \tag{6.12}$$

We shall now show that

$$E_{2,\widetilde{G}}(\widetilde{A}(p_0)) \neq E_{2,\widetilde{A}}(\widetilde{A}(p_0)) = \{t\, e_2\colon\ t \in \mathbb{R}\}. \tag{6.13}$$

For $p \in \mathbb{R}^3$ one can write

$$E_{2,\widetilde{G}}(p) = \{t\,(u(p)e_1 + e_2 + w(p)e_3) : t \in \mathbb{R}\},$$

where $u(p)$ and $w(p)$ are continuous functions on $\mathbb{R}^3$ and $u^2(p) + w^2(p) \le 1$. The differential of $\widetilde{G}$ is given by

$$d\widetilde{G} = \begin{pmatrix} \lambda_1 + \frac{\partial f}{\partial \alpha} & \frac{\partial f}{\partial \beta} & \frac{\partial f}{\partial \gamma} \\ 0 & \lambda_2 & 0 \\ 0 & 0 & \lambda_3 \end{pmatrix}.$$

Since $E_{2,\widetilde{G}}(p)$ is $d\widetilde{G}$-invariant we have

$$C \begin{pmatrix} u(p) \\ 1 \\ w(p) \end{pmatrix} = \begin{pmatrix} \lambda_1 + \frac{\partial f}{\partial \alpha}(\widetilde{G}^{-1}(p)) & \frac{\partial f}{\partial \beta}(\widetilde{G}^{-1}(p)) & \frac{\partial f}{\partial \gamma}(\widetilde{G}^{-1}(p)) \\ 0 & \lambda_2 & 0 \\ 0 & 0 & \lambda_3 \end{pmatrix} \begin{pmatrix} u(\widetilde{G}^{-1}(p)) \\ 1 \\ w(\widetilde{G}^{-1}(p)) \end{pmatrix}.$$

This implies that

$$\begin{aligned} \lambda_2 u(p) = \Big[\lambda_1 + \frac{\partial f}{\partial \alpha}(\widetilde{G}^{-1}(p))\Big] u(\widetilde{G}^{-1}(p)) \\ + \frac{\partial f}{\partial \beta}(\widetilde{G}^{-1}(p)) + \frac{\partial f}{\partial \gamma}(\widetilde{G}^{-1}(p)) w(\widetilde{G}^{-1}(p)) \end{aligned} \tag{6.14}$$

and

$$\lambda_2 w(p) = \lambda_3 w(\widetilde{G}^{-1}(p)). \tag{6.15}$$

It follows from (6.15) that $w(p) = 0$ for all $p \in \mathbb{R}^3$. Thus,

$$E_{2,\widetilde{G}}(p) = \{t\,(u(p)e_1 + e_2) : t \in \mathbb{R}\},$$

where $|u(p)| \leq 1$ and

$$\lambda_2 u(p) = \Big[\lambda_1 + \frac{\partial f}{\partial \alpha}(\widetilde{G}^{-1}(p))\Big]u(\widetilde{G}^{-1}(p)) + \frac{\partial f}{\partial \beta}(\widetilde{G}^{-1}(p)). \tag{6.16}$$

Assume that (6.13) does not hold, i.e., $E_{2,\widetilde{G}}(\widetilde{A}(p_0)) = E_{2,\widetilde{A}}(\widetilde{A}(p_0))$. We will show that in this case the angle between $E_{2,\widetilde{G}}(p)$ and $E_{2,\widetilde{A}}(p)$ will exceed the upper bound given in Statement 3 of Lemma 3.5 at some point (see Figure 10).

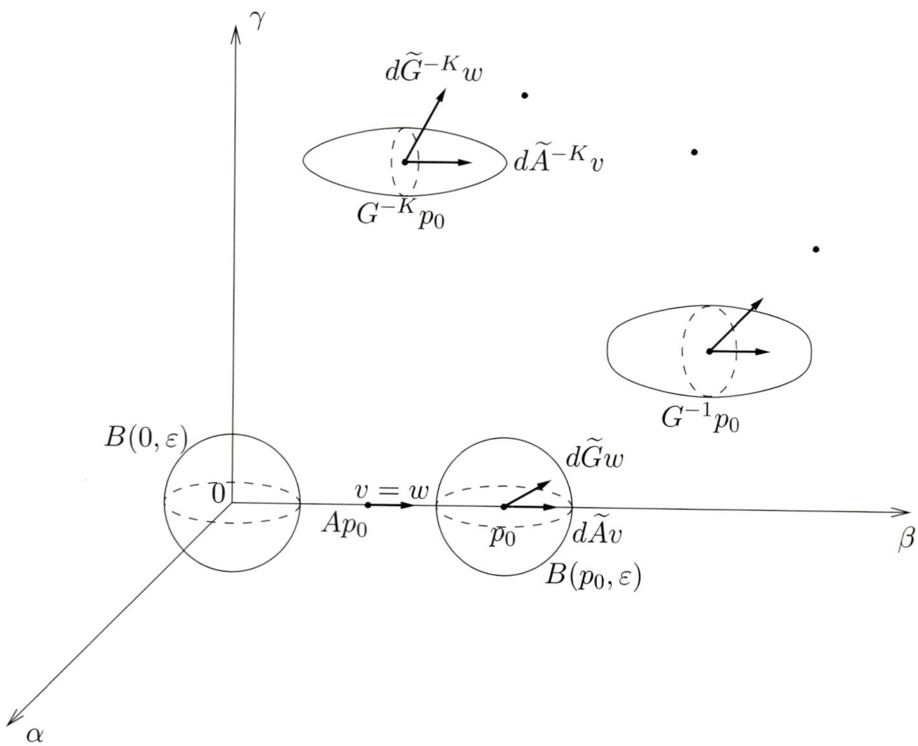


FIGURE 10. To the proof of non-integrability: the vectors $v \in E_{2,\widetilde{A}}(\widetilde{A}p_0)$, $w \in E_{2,\widetilde{G}}(\widetilde{A}p_0)$ and the angle $\angle(d\widetilde{G}^{-K}w, d\widetilde{A}^{-K}v)$ become large.

There exists $q \in B(p_0, \varepsilon)$ such that

$$\widetilde{G}(q) = \widetilde{A}(p_0) = (0, \lambda_2\beta_0, 0).$$

We estimate $|u(q)|$ which gives us the estimate on the angle. Suppose that $q = (\alpha', \beta', \gamma')$. Then

$$\widetilde{G}(q) = (\lambda_1\alpha' + f(\alpha',\ \beta',\ \gamma'), \lambda_2\beta', \lambda_3\gamma').$$

Thus $\gamma' = 0$, $\beta' = \beta_0$, and $\lambda_1\alpha' + f(\alpha', \beta', \gamma') = 0$. It follows that

$$\alpha' = -\lambda_1^{-1} f(\alpha', \beta', \gamma').$$

Therefore,

$$|\alpha'| = \frac{f(\alpha', \beta', \gamma')}{\lambda_1} \le \frac{\delta}{\lambda_1} \le \frac{\varepsilon}{4}.$$

This implies that $q \in B(p_0, \frac{\varepsilon}{3})$. Our assumption means that $u(\widetilde{A}(p_0)) = 0$. We have by equation (6.16) that

$$\Big[\lambda_1 + \frac{\partial f}{\partial \alpha}(q)\Big] u(q) + \frac{\partial f}{\partial \beta}(q) = 0$$

and hence,

$$u(q) = \frac{\frac{\partial f}{\partial \beta}(q)}{\lambda_1 + \frac{\partial f}{\partial \alpha}(q)}.$$

From the definition of the function $f$ it follows that

$$|u(q)| = \frac{|\frac{\partial f}{\partial \beta}(q)|}{|\lambda_1 + \frac{\partial f}{\partial \alpha}(q)|} \ge \frac{\frac{\delta}{4}}{2\lambda_1} = \frac{\delta}{8\lambda_1}.$$

Consider the angles between the subspaces $E_{2,\widetilde{G}}$ and $E_{2,\widetilde{A}}$ at the points $\widetilde{G}^{-1}(q)$, $\widetilde{G}^{-2}(q)$, $\dots$, $\widetilde{G}^{-K}(q)$. Since all these points are outside of $B_\varepsilon(p_0)$ we have

$$\frac{\partial f}{\partial \alpha} = \frac{\partial f}{\partial \beta} = 0$$

at all of them. It follows that

$$\lambda_2 u(\widetilde{G}^{-j}(q)) = \lambda_1 u(\widetilde{G}^{-j-1}(q))$$

and

$$u(\widetilde{G}^{-j-1}(q)) = \frac{\lambda_2}{\lambda_1} u(\widetilde{G}^{-j}(q))$$

for $j = 0, 1, \dots, K-1$. Thus,

$$u(\widetilde{G}^{-K}(q)) = \Big(\frac{\lambda_2}{\lambda_1}\Big)^{K-1} u(\widetilde{G}^{-1}(q)) = \Big(\frac{\lambda_2}{\lambda_1}\Big)^{K} u(q)$$

and

$$|u(\widetilde{G}^{-K}(q))| = \left(\frac{\lambda_2}{\lambda_1}\right)^{K-1} |u(\widetilde{G}^{-1}(q))| \geq \left(\frac{\lambda_2}{\lambda_1}\right)^{K} \frac{\delta}{8\lambda_1}.$$

Since the distribution $E_{2,A}$ is smooth the Hölder exponent in Statement 3 of Lemma 3.5 is one. Therefore, we have the following estimate for the Hölder constant

$$\left(\frac{\lambda_2}{\lambda_1}\right)^{K} \frac{\delta}{8\lambda_1} \leq L.$$

However, this contradicts (6.11).

We now conclude that for all $n \geq 1$,

$$E_{2,\widetilde{G}}(\widetilde{A}^n(p_0)) \neq E_{2,\widetilde{A}}(\widetilde{A}^n(p_0)). \tag{6.17}$$

Indeed, $\widetilde{G} = \widetilde{A}$ near $\widetilde{A}^n(p_0)$, $E_{2,\widetilde{G}}(\widetilde{A}^n(p_0))$ is $\widetilde{G}$-invariant and $E_{2,\widetilde{A}}(\widetilde{A}^n(p_0))$ is $\widetilde{A}$-invariant for any $n$. Therefore, if $E_{2,\widetilde{G}}(\widetilde{A}(p_0)) \neq E_{2,\widetilde{A}}(\widetilde{A}(p_0))$ then $E_{2,\widetilde{G}}(\widetilde{A}^n(p_0)) \neq E_{2,\widetilde{A}}(\widetilde{A}^n(p_0))$.

Since $\widetilde{A}^n p_0 = (0, \lambda_2^{\beta_0}, 0) \to 0$ as $n \to \infty$ we have that $\widetilde{A}^n p_0 \in V_{2,\widetilde{A}}(0)$ for all sufficiently large $n$. For those $n$, the relation (6.17) means that $E_{2,\widetilde{G}}(\widetilde{A}^n(p_0))$ is not tangential to the local smooth manifold $V_{2,\widetilde{A}}(0)$ (since $E_{2,\widetilde{A}}(\widetilde{A}^n(p_0))$ is tangential). In view of (6.12), $V_{2,\widetilde{G}}(0)$ and $V_{2,\widetilde{A}}(0)$ do not coincide near the origin. This implies that $W_{2,\widetilde{G}}(0)$ is not of class $C^N$ and the second statement of the theorem follows.

To prove the third statement let $W_{2,G}(\pi(0))$ be the integral manifold of $E_{2,G}$. It is not of class $C^N$ at $x = \pi(0)$. Denote by

$$S = \{x \in \mathbb{T}^3\colon\ W_{2,G}(x) \text{ is not of class } C^N\}$$

and by

$$I = \{x \in \mathbb{T}^3\colon\ \text{the semi-trajectory } \{G^{-n}(x)\colon\ n \geq 0\} \text{ is dense in } \mathbb{T}^3\}.$$

Since $G$ is topologically conjugate to an Anosov automorphism $A$ it is topologically transitive. Therefore, $I$ is a Baire set. We shall show that $I \subset S$. Otherwise there is $x \in I$ but $x \notin S$. Then $W_{2,G}(x)$ is of class $C^N$ and hence, $V_{2,G}(x) \subset W_{2,G}(x)$. Since these manifolds are $G$-invariant we have $V_{2,G}(G^{-n}(x)) \subset W_{2,G}(G^{-n}(x))$. Furthermore, the size of $V_{2,G}(G^{-n}(x))$ grows as $n$ increases. Since $\{G^{-n}(x)\}$ is dense in $\mathbb{T}^3$ there exists a subsequence $n_i$ such that $G^{-n_i}(x) \to \pi(0)$. Note that $V_{2,G}(x)$ depends continuously on $x$ in the $C^q$ topology. Hence, $V_{2,G}(G^{-n_i}(x)) \to V_{2,G}(\pi(0))$, On the other hand, since the foliation $W_{2,G}$ is continuous, $V_{2,G}(\pi(0)) \subset W_{2,G}(\pi(0))$ by the uniqueness of the limit. This implies that $W_{2,G}(\pi(0))$ is also of class $C^N$ as $V_{2,G}(\pi(0))$. This contradicts the assumption that $x \notin S$. □

## 7. Absolute Continuity

Absolute Continuity is one of the most crucial properties of stable and unstable foliations. It was introduced by Anosov in [1]. Let $W$ be a foliation of $M$ with smooth leaves and $V(x)$, $x \in M$, the local leaf passing through $x$. The absolute continuity property addresses the following question:

> *Let $E \subset B(x,q)$ be a Borel set of positive volume, can the intersection $E \cap V(y)$ have zero Lebesgue measure (with respect to the Riemannian volume on $V(y)$) for almost every $y \in E$?*

If the foliation $W$ is smooth then due to the Fubini theorem, the intersection $B \cap V(y)$ has positive measure for almost all $y \in B(x,q)$. If the foliation is only continuous the absolute continuity property may fail (see Sections 7.4 and 10.3).

**7.1. The holonomy map.** Consider a partially hyperbolic diffeomorphism $f$ of class $C^2$ with a $df$-invariant splitting of the tangent bundle (4.1) satisfying (4.2) and (4.3). Let $W$ be the foliation tangent to the stable distribution $E$.

Fix $x \in M$, $r > 0$ and consider the family of local manifolds

$$\mathcal{L}(x) = \{V(w) : w \in B(x,r)\}. \tag{7.1}$$

Choose two local disks $D^1$ and $D^2$, which are transverse to the family $\mathcal{L}(x)$, and define the *holonomy map* $\pi = \pi(x,W)\colon D^1 \to D^2$ (generated by the family of local manifolds) by setting

$$\pi(y) = D^2 \cap V(w) \text{ if } y = D^1 \cap V(w), \quad w \in B(x,r).$$

The holonomy map $\pi$ is a homeomorphism onto its image (see Figure 11).

A measurable invertible transformation $T\colon X \to Y$ of measurable spaces $(X,\nu)$ and $(Y,\mu)$ is said to be *absolutely continuous* if the measure $\mu$ is absolutely continuous with respect to the measure $T_*\nu$. In this case one defines the *Jacobian* $\operatorname{Jac}(T)(x)$ of $T$ at a point $x \in X$ (specified by the measures $\nu$ and $\mu$) to be the Radon–Nikodym derivative $\frac{d\mu}{d(T_*\nu)}$. If $X$ is a metric space then for $\mu$-almost every $x \in X$ one has

$$\operatorname{Jac}(T)(x) = \lim_{r\to 0} \frac{\mu(T(B(x,r)))}{\nu(B(x,r))}. \tag{7.2}$$

Let $m$ denote the Riemannian volume. Given a submanifold $D$ in $M$, let $m_D$ be the Riemannian volume on $D$ induced by the restriction of the Riemannian metric to $D$. Finally, let $\operatorname{Jac}(\pi)(y)$ be the Jacobian of the holonomy map $\pi$ at the point $y \in D^1$ specified by the measures $m_{D^1}$ and $m_{D^2}$.

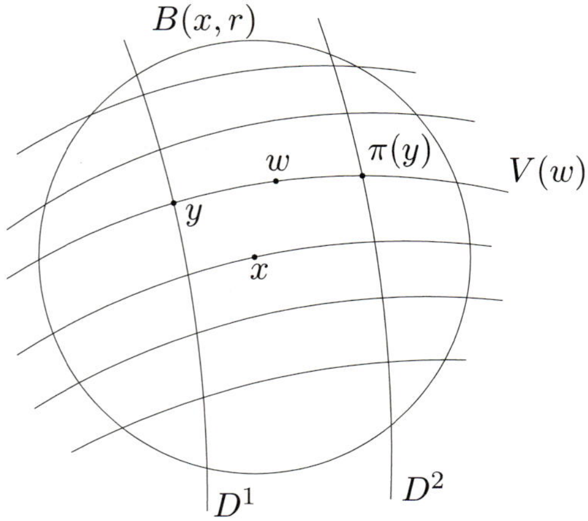


FIGURE 11. The holonomy map.

**Theorem 7.1** (Brin and Pesin, [13], Pugh and Shub, [40, 41], see also [3]). *Let $f$ be a partially hyperbolic $C^2$ diffeomorphism of a compact smooth manifold $M$. Given $x \in M$ and two transverse disks $D^1$ and $D^2$ to the family $\mathfrak{L}(x)$ of local stable manifolds $V(y)$, $y \in B(x,r)$, the holonomy map $\pi$ is absolutely continuous (with respect to the measures $m_{D^1}$ and $m_{D^2}$) and the Jacobian $Jac(\pi)$ is bounded from above and bounded away from zero.*

**Remark 7.2.** One can show (see [3], Chapter 4.4) that the Jacobian $\mathrm{Jac}(\pi)$ can be computed by the following formula: for any $y \in D^1$,

$$\mathrm{Jac}(\pi)(y) = \prod_{n=0}^{\infty} \frac{\mathrm{Jac}(d_{f^n(\pi(y))} f^{-1} | T_{f^n(\pi(y))} f^n(D^2))}{\mathrm{Jac}(d_{f^n(y)} f^{-1} | T_{f^n(y)} f^n(D^1))} \tag{7.3}$$

(in particular, the infinite product on the right hand-side converges).

*Sketch of the proof of the theorem.* Let us first outline our approach. To estimate the Jacobian $\mathrm{Jac}(\pi)$ choose a small open set $A \subset D^1$ and let $B = \pi(A) \subset D^2$. We need to compare the measures $\nu_{D^1}(A)$ and $\nu_{D^2}(B)$. Consider the images $f^m(A)$ and $f^m(B)$, $m > 0$ which are smooth submanifolds of $M$. When $m$ increases the sets $A$ and $B$ get stretched in the unstable direction and may get stretched and/or shrunk in the central direction. The latter may occur with at most an exponential rate $\gamma$ with some $\lambda < \gamma < \min\{1, \mu\}$. On the other hand, the distance between the sets $f^m(A)$ and $f^m(B)$ gets small exponentially with a rate $\lambda'$ where $\lambda < \lambda' < \gamma$.

We then cover the set $f^m(A)$ by open sets $A_i$ of size $\gamma^m$ such that the multiplicity $K$ of the cover is finite and depends only on the dimension of $D^1$.

The distance between the sets $A_i$ and $B_i = \pi(A_i)$ is exponentially smaller than their sizes and therefore, the ratio of their measures is bounded from above and away from zero (uniformly in $i$ and $m$). So is the ratio of the measures of the sets $f^m(A)$ and $f^m(B)$ (note that the sets $B_i$ form a cover of $f^m(B)$ of multiplicity $K$).

To return back to the measures $\nu_{D^1}(A)$ we use the well-known formulae

$$\nu_{D^1}(A) = \int_{f^m(A)} \operatorname{Jac}(df^{-m}|T_y f^m(D^1))(y)\,d\nu_{f^m(D^1)}(y),$$

$$\nu_{D^2}(B) = \int_{f^m(B)} \operatorname{Jac}(df^{-m}|T_y f^m(D^2))(y)\,d\nu_{f^m(D^2)}(y).$$

It remains to estimate the ratio of the Jacobians $\operatorname{Jac}(df^{-m}|T_y f^m(D^1))(y)$ and $\operatorname{Jac}(df^{-m}|T_y f^m(D^2))(\pi(y))$ for $y \in f^m(A)$.

We split the proof into several steps.

**Step I.** Fix a point $w \in B(x,r)$ and let $y^i = V(w) \cap D^i$ for $i = 1$, 2. Fix also a number $\gamma$ such that

$$\lambda < \gamma < \min\{1, \mu\}. \tag{7.4}$$

Choose $q > 0$ and set for $i = 1$, 2 and $m \geq 0$,

$$D^i_m = f^m(D^i), \quad w_m = f^m(w), \quad y^i_m = f^m(y_i), \quad q_m = q\gamma^m.$$

Note that $D^i_0 = D^i$, $w_0 = w$, $y^i_0 = y^i$ and $q_0 = q$. For sufficiently small $q > 0$ there exists an open neighborhood $D^i_m(w,q) \subset D^i_m$ of the point $y^i_m$ such that

$$D^i_m(w,q) = \exp_{w_m}\{(\psi^i_m(v), v) \colon v \in B(q_m)\},$$

where $B(q_m) \subset F(w_m)$ is the ball centered at the origin of radius $q_m$ (see Figure 12).

Furthermore, there are numbers $C > 0$ and $\lambda < \lambda' < \gamma$ such that for $i = 1, 2$ and $m \geq 0$,

$$\max_{v \in B(q_m)} \|d\psi^i_m(v)\| \leq K, \qquad \max_{v \in B(q_m)} \|d^2\psi^i_m(v)\| \leq K, \tag{7.5}$$

$$\rho(D^1_m(w,q), D^2_m(w,q)) \leq K(\lambda')^m. \tag{7.6}$$

It follows that for $n = 0, \ldots, m$,

$$f^{-1}(D^i_n(w,q)) \subset D^i_{n-1}(w,q),\ i = 1,\ 2 \tag{7.7}$$

provided that $D^1$ and $D^2$ are chosen sufficiently close to each other (this does not depend on $m$). Properties (7.5) and (7.6) mean that $D^i_m(w,q)$ is roughly a ball in the direction transversal to $E(w_m)$ whose size decays exponentially with the rate $\gamma^m$ while the distance between $D^1_m(w,q)$ and $D^2_m(w,q)$ decays exponentially with the rate $(\lambda')^m$.

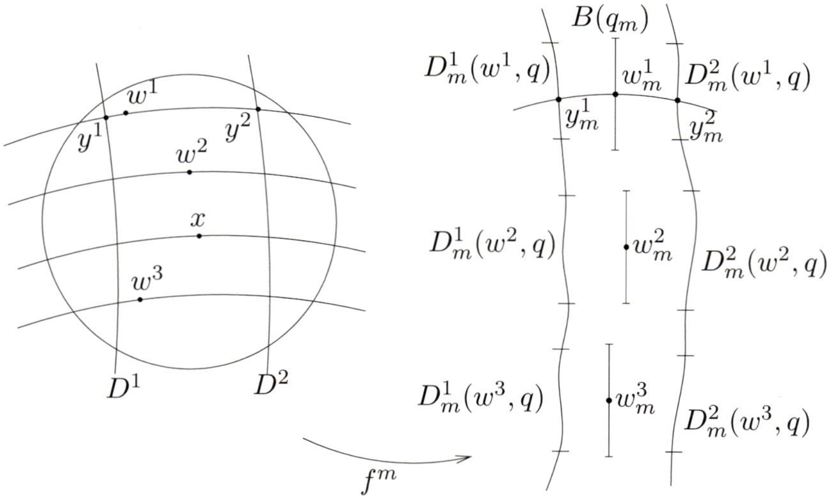


FIGURE 12. To the proof of Theorem 7.1.

This allows us to compare the Riemannian volumes $\nu_{D^1_m}|D^1_m(w,q)$ and $\nu_{D^2_m}|D^2_m(w,q)$ for sufficiently large $m$.

**Lemma 7.3.** *There exists $K_1 > 0$ such that the following holds: for any $m > 0$ there exists $q_0 = q_0(m) > 0$ such that for any $0 < q \le q_0$,*

$$\left| \frac{\nu_{D^1_m}(D^1_m(w,q))}{\nu_{D^2_m}(D^2_m(w,q))} - 1 \right| \le K_1.$$

We need the following Covering Lemma.

**Lemma 7.4.** *For any $m > 0$ there are points $w^j \in B(x,r)$, $j = 1, \dots, p = p(m)$ and a number $q = q(m) > 0$ such that for $i = 1, 2$ the sets $D^i_m(w^j,q)$ form an open cover of the set $D^i_m$ of finite multiplicity which depends only on the dimension of $D^i$.*

**Step II.** We now compute the measures that are the pullbacks under $f^{-m}$ of the measures $\nu_{D^i_m}|D^i_m(w,q)$ for $i = 1, 2$. Note that for every $w \in B(x,r)$ the points $y^1_m$ and $y^2_m$ lie on the local manifold $V(w_m)$. For any $z \in f^{-m}(D^i_m(w,q))$ (with $i = 1$ or $i = 2$) set $z_m = f^m(z)$ and let

$$J^i(z,m) = \operatorname{Jac}(d_{z_m} f^{-m} | T_{z_m} D^i_m(w,q)).$$

We need the following lemma.

**Lemma 7.5.** *There exist $K_2 > 0$ and $m_1 > 0$ such that for every $w \in B(x,r)$ and $m \geq m_1$ one can find $q = q(m)$ such that*

$$\left| \frac{J^2(y_m^2, m)}{J^1(y_m^1, m)} - 1 \right| \leq K_2,$$

*and if $z \in D_0^1(w,q)$ then*

$$\left| \frac{J^1(z_m, m)}{J^1(y_m^1, m)} - 1 \right| \leq K_2.$$

*Proof of the lemma.* For any $0 < n \leq m$ and $z \in D_0^1(w,q)$ let us set

$$V(n,z) = T_{z_n} D_n^1(w,q).$$

We transport parallelly the space $V(n,z) \subset T_{z_n}M$ along the geodesic that connects the points $z_n$ and $y_n^1$ (this geodesic is uniquely defined since these points are sufficiently close to each other) to obtain a new subspace $\widetilde{V}(n,z) \subset T_{y_n^1}M$. We have

$$\begin{aligned} &\left|\operatorname{Jac}(d_{z_n} f^{-1}|V(n,z)) - \operatorname{Jac}(d_{y_n^1} f^{-1}|V(n,y^1))\right| \\ &\leq \left|\operatorname{Jac}(d_{z_n} f^{-1}|V(n,z)) - \operatorname{Jac}(d_{y_n^1} f^{-1}|\widetilde{V}(n,z))\right| \\ &+ \left|\operatorname{Jac}(d_{y_n^1} f^{-1}|\widetilde{V}(n,z)) - \operatorname{Jac}(d_{y_n^1} f^{-1}|V(n,y^1))\right|. \end{aligned} \tag{7.8}$$

Since $f$ is of class $C^2$,

$$|\operatorname{Jac}(d_{z_n} f^{-1}|V(n,z)) - \operatorname{Jac}(d_{y_n^1} f^{-1}|\widetilde{V}(n,z))| \leq C_1 \rho(z_n, y_n^1) \tag{7.9}$$

and

$$\begin{aligned} &|\operatorname{Jac}(d_{y_n^1} f^{-1}|\widetilde{V}(n,z)) - \operatorname{Jac}(d_{y_n^1} f^{-1}|V(n,y^1))| \\ &\qquad \leq C_2 d(\widetilde{V}(n,z), V(n,y^1)), \end{aligned} \tag{7.10}$$

where $C_1 > 0$ and $C_2 > 0$ are constants (here $d$ is the distance in the Grassmannian bundle of $TM$ generated by the Riemannian metric). By (7.5),

$$d(\widetilde{V}(n,z), V(n,y^1)) \leq C_3 \rho(z_n, y^1), \tag{7.11}$$

where $C_3 > 0$ is a constant. It follows from (7.8)–(7.11) that

$$\begin{aligned} &|\operatorname{Jac}(d_{z_n} f^{-1}|V(n,z)) - \operatorname{Jac}(d_{y_n^1} f^{-1}|V(n,y^1))| \\ &\qquad \leq C_4 \rho(z_n, y_n^1) \leq C_5 \mu^k \rho(z, y_1), \end{aligned}$$

where $C_4 > 0$ and $C_5 > 0$ are constants. Note that for any $0 < n \leq m$ and $z \in D_0^1(w,q)$,

$$C_6^{-1} \leq |\operatorname{Jac}(d_{z_n} f^{-1}|V(n,z))| \leq C_6,$$

where $C_6 > 0$ is a constant. We obtain

$$\begin{aligned}\frac{J^1(z_m, m)}{J^1(y^1_m, m)} &= \prod_{n=1}^{m} \frac{\operatorname{Jac}(d_{z_n} f^{-1}|V(n,z))}{\operatorname{Jac}(d_{y^1_n} f^{-1}|V(n,y^1))} \\ &= \exp \sum_{n=1}^{m} \log \frac{\operatorname{Jac}(d_{z_n} f^{-1}|V(n,z))}{\operatorname{Jac}(d_{y^1_n} f^{-1}|V(n,y^1))} \\ &\leq \exp \sum_{n=1}^{m} \left( \frac{\operatorname{Jac}(d_{z_n} f^{-1}|V(n,z))}{\operatorname{Jac}(d_{y^1_n} f^{-1}|V(n,y^1))} - 1 \right) \\ &\leq \exp \left( C_5 C_6 \rho(z, y^1) \sum_{n=1}^{m} \mu^n \right).\end{aligned}$$

For a fixed $m$ the last expression can be made arbitrarily close to 1 by choosing $q$ sufficiently small. This proves the second inequality. The first inequality can be proven in a similar fashion if one observes that by the Hölder continuity of the distribution $E(x)$ (see Theorem 3.8),

$$d(\widetilde{V}(n,z), V(n,y^1)) \leq C_7 \rho(z_n, y^1)^\alpha,$$

where $C_7 > 0$ is a constant, $\alpha > 0$ is the Hölder exponent and $\widetilde{V}(n,z)$ is the parallel transport of $V(n,z)$ along the geodesic connecting $y^1$ and $y^2$. □

Lemma 7.5 allows one to compare the measures of the preimages under $f^{-m}$ of $D^1_m(w,q)$ and $D^2_m(w,q)$. More precisely, the following statement holds.

**Lemma 7.6.** *There exist $K_3 > 0$ and $m_2 > 0$ such that if $w \in B(x,r)$ and $m \geq m_2$, then one can find $q = q(m)$ such that*

$$\left| \frac{\nu_{D^1}(f^{-m}(D^1_m(w,q)))}{\nu_{D^2}(f^{-m}(D^2_m(w,q)))} - 1 \right| \leq K_3.$$

*Proof of the lemma.* For $i = 1$, 2 we have

$$\begin{aligned}\nu_{D^i}(f^{-m}(D^i_m(w,q))) &= \int_{D^i_m(w,q)} J^i(z,m)\, d\nu_{D^i_m}(z) \\ &= J^i(z^i_m, m)\nu_{D^i_m}(D^i_m(w,q)),\end{aligned}$$

where $z^i_m \in D^i_m(w,q)$ are some points. It follows from the assumptions of the lemma, (7.7), and Lemmas 7.3 and 7.5 that for sufficiently large $m$ and small $q$,

$$\begin{aligned}
&\left|\frac{\nu_{D^1}(f^{-m}(D^1_m(w,q)))}{\nu_{D^2}(f^{-m}(D^2_m(w,q)))}-1\right| \\
&\le \left|\frac{J^1(z^1_m,m)}{J^2(z^2_m,m)}-1\right| \times \frac{\nu_{D^1_m}(D^1_m(w,q))}{\nu_{D^2_m}(D^2_m(w,q))} + \left|\frac{\nu_{D^1_m}(D^1_m(w,q))}{\nu_{D^2_m}(D^2_m(w,q))}-1\right| \\
&\le \left|\frac{J^1(z^1_m,m)}{J^1(y^1_m,m)}\cdot\frac{J^2(y^2_m,m)}{J^2(z^2_m,m)}\cdot\frac{J^1(y^1_m,m)}{J^2(y^2_m,m)}-1\right| \times (1+K_1)+K_1\Delta \\
&\le ((1+K_2)^3-1)(1+K_1)+K_1 \le K_3.
\end{aligned}$$

The lemma follows. □

**Step III.** Given $m>0$, choose points $w_j \in B(x,r)$ and a number $q=q(m)$ as in Lemma 7.4. Consider the sets

$$\hat{D}^1_m = \bigcup_{j=1}^p D^1_m(w_j,q), \quad \hat{D}^2_m = \bigcup_{j=1}^p D^2_m(w_j,q). \tag{7.12}$$

Note that $\hat{D}^i_m \supset f^m(D^i)$. We wish to compare the measures $\nu_{D^1}|f^{-m}(\hat{D}^1_m)$ and $\nu_{D^2}|f^{-m}(\hat{D}^2_m)$. Let $L_i$ be the multiplicities of the covers $\{D^i_m(w_j,q)\}$, $i=1$, 2. Observe that the cover of the set $D^i$ by the sets $\{f^{-m}(D^i_m(w_j,q))\}$ is also of multiplicity $L_i$. Set $L=\max\{L_1,L_2\}$. We have

$$\begin{aligned}
\frac{1}{L}\sum_{j=1}^p \nu_{D^i}(f^{-m}(D^i_m(w_j,q))) &\le \nu_{D^i}(f^{-m}(\hat{D}^i_m)) \\
&\le \sum_{j=1}^p \nu_{D^i}(f^{-m}(D^i_m(w_j,q))).
\end{aligned}$$

It follows from Lemma 7.6 that

$$\begin{aligned}
\frac{\nu_{D^1}(f^{-m}(\hat{D}^1_m))}{\nu_{D^2}(f^{-m}(\hat{D}^2_m))} &\le L\frac{\sum_{j=1}^p \nu_{D^1}(f^{-m}(D^1_m(w_j,q)))}{\sum_{j=1}^p \nu_{D^2}(f^{-m}(D^2_m(w_j,q)))} \\
&\le L\max\left\{\frac{\nu_{D^1}(f^{-m}(D^1_m(w_j,q)))}{\nu_{D^2}(f^{-m}(D^2_m(w_j,q)))} : j=1,\dots,p\right\} \\
&\le L\,(1+K_3)\,,
\end{aligned}$$

with a similar bound for the inverse ratio. We conclude that

$$K_4^{-1} \le \frac{\nu_{D^1}(f^{-m}(\hat{D}^1_m))}{\nu_{D^2}(f^{-m}(\hat{D}^2_m))} \le K_4, \tag{7.13}$$

where $K_4>0$ is a constant independent of $m$.

**Step IV.** Without loss of generality we may assume that $\nu_{D^i}(D^i) > 0$. Given $\beta > 0$, denote by $U^i_\beta$ the $\beta$-neighborhood of the set $D^i$ for $i = 1, 2$. We need the following lemma from measure theory.

**Lemma 7.7.** *There exists* $\beta_0 > 0$ *such that for every* $0 < \beta \leq \beta_0$ *and* $i = 1, 2$,

$$\frac{\nu_{D^i}(D^i)}{\nu_{D^i}(U^i_\beta)} \geq \frac{1}{2}.$$

For any $\beta > 0$ and any sufficiently large $m > 0$ we have $D^i \subset f^{-m}(\hat{D}^i_m) \subset U^i_\beta$ for $i = 1, 2$. It follows from Lemma 7.7 and (7.13) that

$$\frac{1}{2K_4} \leq \frac{\nu_{D^1}(D^1)}{\nu_{D^2}(D^2)} \leq 2K_4.$$

We emphasize that the constant $K_4$ does not depend on the size of the transversals $D^1$ and $D^2$ but only on their dimension. In particular, for each $y \in D^1$,

$$\frac{1}{2K_4} \leq \frac{\nu_{D^2}(\pi(B^1(y,r)))}{\nu_{D^1}(B^1(y,r))} \leq 2K_4,$$

where $B^1(y,r) \subset D^1$ is the ball of radius $r$ centered at $y$. Letting $r \to 0$ we conclude from (7.2) that

$$\frac{1}{2K_4} \leq \operatorname{Jac}(\pi)(y) \leq 2K_4$$

for $\nu_{D^1}$-almost every $y \in D^1$. Therefore, the Jacobian is bounded from above and bounded away from zero. This completes the proof of the theorem. $\square$

**7.2. Absolute continuity of local manifolds.** Fix $x \in M$, $r > 0$ and set

$$Q(x) = \bigcup_{w \in B(x,r)} V(w),$$

where $B(x,r)$ is the ball centered at $x$ of radius $r$. Consider the measurable partition $\xi$ of $Q(x)$ into local manifolds.

Let $\nu$ be a smooth $f$-invariant probability measure on $M$. It generates conditional measures $\nu_V(w)$ on $V(w)$, $w \in B(x,r)$. The factor space $Q(x)/\xi$ can be identified with an open transverse disk $D$. We denote by $\hat{\nu}_D$ the factor measure generated by the partition $\xi$ (it is supported on the set $D$), by $m_V(w)$ the Riemannian volume on $V(w)$, and by $m_D$ the Riemannian volume on $D$.

The following theorem shows that the foliation $W$ satisfies the Fubini theorem.

**Theorem 7.8.** *The following statements hold:*

(1) *the measures $\nu_V(w)$ and $m_V(w)$ are equivalent for $\nu$-almost every $w \in B(x, r)$;*
(2) *the factor measure $\hat{\nu}_D$ is equivalent to the measure $m_D$.*

*Proof.* Consider a family of smooth disjoint transversals $D(w)$, $w \in B(x, r)$, to the family of local manifolds $\mathfrak{L}(x)$. They generate a partition $\eta$ of the ball $B(x, r)$. Let us denote by $m_D(w)$ the Riemannian volume on $D(w)$, by $\nu_D(w)$ the system of conditional measures generated by the measure $\nu$ and partition $\eta$, and by $\hat{\nu}_V$ the factor measure generated by the partition $\eta$ and the measure $\nu$. One can see that the factor space $B(x, r)/\eta$ can be identified with the local manifold $V(w)$ for some $w \in B(x, r)$. Since the partition $\eta$ is smooth there exist continuous functions $g(w, z)$ and $h(w)$ such that for $w \in B(x, r)$ and $z \in D(w)$,

$$d\nu_D(w)(z) = g(w, z)\, dm_D(w)(z), \quad d\hat{\nu}_V(w) = h(w)\, d\nu_V(w).$$

Let $X$ be a Borel subset of positive $\nu$-measure. We have

$$\begin{aligned}
\nu(X) &= \int_{B(x,r)/\eta} \int_{D(w)} \chi_X(w, z)\, d\nu_D(w)(z)\, d\hat{\nu}_V(w) \\
&= \int_{B(x,r)/\eta} \int_{D(w)} \chi_X(w, z) g(w, z)\, dm_D(w)(z)\, d\hat{\nu}_V(w) \\
&= \int_{B(x,r)/\eta} \int_{D(w_0)} \chi_X(w, z) g(w, z) \mathrm{Jac}(\pi_{w_0 w})(y)\, dm_D(w_0)(y)\, d\hat{\nu}_V(w),
\end{aligned}$$

where $\chi_X(w, z)$ is the characteristic function of the set $X$ at the point $z \in D(w)$, $\pi_{w_0 w}$ the holonomy map between the transversals $D(w_0)$ and $D(w)$, $z = \pi_{w_0 w}(y)$, and $\mathrm{Jac}(\pi_{w_0 w})(y)$ is the Jacobian of the map $\pi_{w_0 w}$ at $y$. Applying Fubini's theorem we obtain

$$\begin{aligned}
\nu(X) &= \int_{D(w_0)} \int_{B(x,r)/\eta} \chi_X(w, z) g(w, z) \mathrm{Jac}(\pi_{w_0 w})(y)\, d\hat{\nu}_V(w)\, dm_D(w_0)(y) \\
&= \int_{D(w_0)} \int_{V(w)} \chi_X(w, z) g(w, z) \mathrm{Jac}(\pi_{w_0 w})(y) h(w)\, dm_V(w)\, dm_D(w_0)(y).
\end{aligned}$$

This implies the desired result. □

As an immediate consequence we obtain that

$$d\nu_V(w)(y) = \kappa(w, y) dm_V(w)(y)$$

for every $w \in B(x,r)$ and $y \in V(w)$. The function $\kappa(w,y)$ is continuous and satisfies the following homological equation

$$\kappa(f(w), f(y)) = \frac{\operatorname{Jac}(df|E(y))}{\operatorname{Jac}(df|E(w))}\,\kappa(w,y).$$

It follows that

$$\kappa(w,y) = \prod_{i=0}^{\infty} \frac{\operatorname{Jac}(df|E(f^i(y)))}{\operatorname{Jac}(df|E(f^i(w)))}.$$

It is easy to see that by the Hölder continuity of the distribution $E$ (see Theorem 3.8), the infinite product converges.

**7.3. Ergodicity of Anosov maps.** Let $f$ be a $C^2$ Anosov diffeomorphism of a smooth compact connected Riemannian manifold $M$ and $TM = E^s \oplus E^u$ the $df$-invariant decomposition of the tangent bundle such that, for every $x \in M$,

$$\|df|E^s(x)\| \le \lambda, \quad \|df|E^u(x)^{-1}\|^{-1} \ge \mu,$$

with $0 < \lambda < 1 < \mu$. Using the absolute continuity of the associated foliations $W^s$ and $W^u$ of $M$, one can prove the following ergodicity result.

**Theorem 7.9.** *If $f$ preserves the Riemannian volume, then $f$ is ergodic.*

*Proof.* We follow an approach which is an elaboration of the Hopf argument (see [2]). Let $\varphi : M \to \mathbb{R}$ be a continuous function. According to the Birkhoff ergodic theorem, the three functions

$$\begin{aligned}
&\varphi^+(x) = \lim_{n\to+\infty} \frac{1}{n}\sum_{k=0}^{n-1} \varphi(f^k(x)), \quad \varphi^-(x) = \lim_{n\to-\infty} \frac{1}{|n|}\sum_{k=0}^{n+1} \varphi(f^k(x)),\\
&\bar\varphi(x) = \lim_{n\to+\infty} \frac{1}{2n+1}\sum_{k=-n}^{n} \varphi(f^k(x))
\end{aligned} \tag{7.14}$$

exist and are equal for almost every $x \in M$. These functions are dense in the space of invariant measurable functions on $M$, since continuous functions are dense in the space of measurable functions on $M$. Let $M_0$ be the conull set in $M$ on which $\varphi^+$, $\varphi^-$ and $\bar\varphi$ exist and are equal. For any $x \in M_0$ and every point $z$ on the local stable manifold $V^s(x)$ we have $|\varphi(f^k(x)) - \varphi(f^k(z))| \to 0$ as $k \to +\infty$. Hence, $\varphi^+(x) = \varphi^+(z)$. By absolute continuity, for almost every $x \in M$, almost every point $z \in V^s(x)$ belongs to $M_0$ and the set

$$\bigcup_{z\in V^s(x),\, z\notin M_0} V^u(z)$$

has measure zero. Hence, for almost every $y$ close enough to $x$, the point $z = V^s(x) \cap V^u(y)$ lies in $M_0$, and thus,

$$\bar{\varphi}(y) = \varphi^-(y) = \varphi^-(z) = \bar{\varphi}(z) = \bar{\varphi}(x).$$

Therefore $\bar{\varphi}$ is constant almost everywhere on a neighborhood of $x$. Since any two points of $M$ can be joined by a path contained in a finite union of such neighborhoods, $\bar{\varphi}$ is constant almost everywhere on $M$. □

**7.4. An example of a non-absolutely continuous foliation.** We describe an example due to Katok of a non-absolutely continuous foliation (see [32] for a similar example). Consider a hyperbolic automorphism $A$ of the torus $\mathbb{T}^2$. There is a family $\{f_t : t \in S^1\}$ of diffeomorphisms preserving the area $m$ and satisfying the following conditions:

1. $f_t$ is a small perturbation of $A$ for every $t \in S^1$;
2. $f_t$ depends smoothly on $t$;
3. the function $h(t) = h_m(f_t)$ is strongly decreasing in a small neighborhood of $t = 0$ (here $h_m(f_t)$ is the metric entropy of the diffeomorphism $f_t$).

Consider the diffeomorphism $F\colon \mathbb{T}^2 \times S^1 \to \mathbb{T}^2 \times S^1$ by $F(x,t) = (f_t(x), t)$. Since $f_t$ is sufficiently close to $A$, they are conjugate via a Hölder homeomorphism $g_t$, i.e., $f_t = g_t \circ A \circ g_t^{-1}$. Given $x \in \mathbb{T}^2$, consider the set $H(x) = \{g_t(x) \colon t \in S^1\}$. It is diffeomorphic to the circle $S^1$ and the collection of these sets forms an $F$-invariant foliation $H$ of $\mathbb{T}^2 \times S^1 = \mathbb{T}^3$. Note that $H(x)$ depends Hölder continuously on $x$. However, the holonomy maps associated with the foliation $H$ are not absolutely continuous. To see this consider the holonomy map

$$\pi_{t_1,t_2} \colon \mathbb{T}^2 \times \{t_1\} \to \mathbb{T}^2 \times \{t_2\}.$$

We have that

$$\pi_{0,t}(x,t) = (g_t(x), t).$$

Since the entropy function $h(t)$ is continuous and strongly decreasing for small $t$ the map $g_t$ is not absolutely continuous for all small $t$, and the same happens with the map $\pi_{0,t}$.

One can show that the foliation $H$ has the following "pathological" property: there is a set of full volume which intersects almost every leaf of the foliation at a set of zero length (one can show that indeed, it intersects almost every leaf at exactly one point, see [44]). This phenomenon is known as "Fubini's nightmare" since Fubini theorem fails with respect to this foliation in the strongest possible way. See Section 8.3 for more information.

# 8. Accessibility and Stable Accessibility

**8.1. The accessibility property.** Unlike volume-preserving Anosov diffeomorphisms partially hyperbolic (in the narrow sense) volume-preserving diffeomorphisms may not be topologically transitive nor ergodic. The simple example that illustrates this is the direct product of a volume-preserving Anosov diffeomorphism and the identity map. Note that in this example the stable and unstable foliations, $W^s$ and $W^u$, are jointly integrable and the leaves of the corresponding foliation are the components of topological transitivity as well as ergodic components. Therefore, one may expect that non-integrability of the pair of foliations $(W^s, W^u)$ would guarantee (or lead to) topological transitivity and/or ergodicity. Since the foliations $W^s$ and $W^u$ are not in general, smooth one cannot use the Lie bracket techniques and apply Frobenius' theorem to establish integrability (or non-integrability). Accessibility is designed to study integrability of the pair of foliations $(W^s, W^u)$ when they are not smooth.

Let $f$ be a partially hyperbolic (in the narrow sense) diffeomorphism of a compact Riemannian manifold $M$.

Two points $p, q \in M$ are called *accessible* , if there are points $z_0 = p, z_1, \dots, z_{\ell-1}, z_\ell = q$, $z_i \in M$, such that $z_i \in V^\alpha(z_{i-1})$ for $i = 1, \dots, \ell$ and $\alpha = s$ or $u$ (see Figure 13). The collection of points $z_0, z_1, \dots, z_\ell$ is called the *us-path* connecting $p$ and $q$ and is denoted by $[p, q]_f = [p, q] = [z_0, z_1, \dots, z_\ell]$. Note that there is an actual path which consists of pieces of smooth curves and connects $p$ and $q$. The curves lie on local stable or unstable manifolds and have the points $z_i$ as endpoints.

Accessibility is an equivalence relation and the collection of points accessible from a given point $p$ is called the *accessibility class* of $p$.

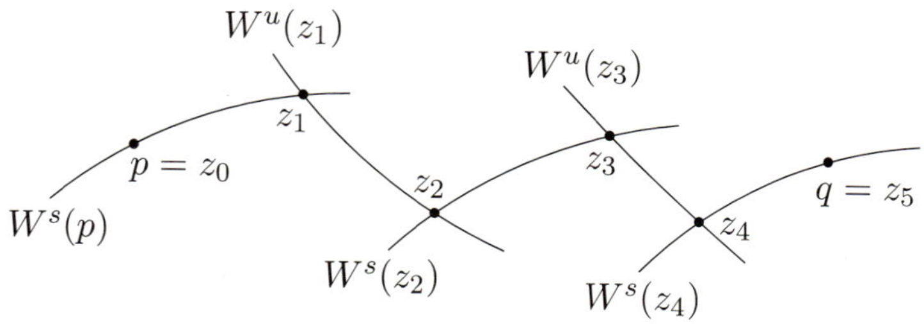


FIGURE 13. A *us*-path connecting points $p$ and $q$.

A diffeomorphism $f$ is said to have the *accessibility property* if the accessibility class of any point is the whole manifold $M$, or, in other words, if any two points are accessible.

If $f$ has the accessibility property then the distribution $E^s \oplus E^u$ is not integrable (and therefore, the stable and unstable foliations, $W^s$ and $W^u$, are not jointly integrable). Otherwise, the accessibility class of any $p \in M$ would be the leaf of the corresponding foliation passing through $p$.

There is a weaker version of accessibility which provides a useful tool in studying topological transitivity of $f$. Given $\varepsilon > 0$, we say that $f$ is $\varepsilon$-*accessible* if for every open ball $B$ of radius $\varepsilon$ the union of accessibility classes passing through $B$ is $M$. An equivalent requirement is that the accessibility class of any point should enter every open ball of radius $\varepsilon$. Clearly, if $f$ is accessible then it is $\varepsilon$-accessible for any $\varepsilon$.

**Proposition 8.1.** *Assume that a partially hyperbolic diffeomorphism $f$ has the accessibility property. Then for every $\varepsilon > 0$*

(1) *there exist $\ell > 0$ and $R > 0$ such that for any $p, q \in M$ one can find a us-path that starts at $p$, ends within distance $\varepsilon/2$ of $q$, and has at most $\ell$ legs, each of them with length at most $R$;*
(2) *there exists a neighborhood $\mathfrak{U}$ of $f$ in the space* $\mathrm{Diff}^2(M)$ *such that every $g \in \mathfrak{U}$ is $\varepsilon$-accessible.*

*Proof.* Let $q_1, \dots, q_N$ be an $\varepsilon/4$-net in $M$. Fix $p \in M$. For each $q_k$, choose a *us*-path from $p$ to $q_k$; let $\ell(p,k)$ be the number of legs and $R(p,k)$ the length of the longest leg in this path. Set

$$R(p) = \max_k R(p,k) \quad \text{and} \quad \ell(p) = \max_k \ell(p,k).$$

By continuity of the foliations $W^u$ and $W^s$, every point $p \in M$ has a neighborhood $U(p)$ such that, for each $k$, any point in $U(p)$ is joined to some point in $B(q_k, \varepsilon/4)$ by a *us*-path which has at most $\ell(p)$ legs each of length at most $2R(p)$. The sets $\{U(p)\}$ form an open cover of $M$. Let $\{U(p_1), \dots, U(p_m)\}$ be a finite subcover. Then $R = \max_i 2R(p_i)$ and $\ell = \max_i \ell(p_i)$ satisfy the first statement.

The second statement follows from the first one and the continuous dependence of the leaves of $W^u_g$ and $W^s_g$ on $g$. □

**8.2. Accessibility and topological transitivity.** Given $\varepsilon > 0$ we say that an orbit is $\varepsilon$-*dense* if the points of the orbit form an $\varepsilon$-net. Clearly, a trajectory $\{f^n(x)\}_{n\in\mathbb{Z}}$ is everywhere dense in $M$ if and only if it is $\varepsilon$-dense for every $\varepsilon > 0$.

We say that a point $x \in M$ is *forward* (respectively, *backward*) *recurrent* if for any $r > 0$ there exists $n > 0$ (respectively, $n < 0$) such that $f^n(x) \in B(x,r)$. If a point $x$ is forward (respectively, backward) recurrent then for any $r > 0$ there exists a sequence $n_k \to +\infty$ (respectively, $n_k \to -\infty$) such that $f^{n_k}(x) \in B(x,r)$.

**Theorem 8.2** (Burns, Dolgopyat and Pesin, [14]). *If a partially hyperbolic (in the narrow sense) diffeomorphism $f$ is $\varepsilon$-accessible and preserves a smooth measure then almost every orbit of $f$ is $\varepsilon$-dense.*

*Proof.* It suffices to show that if $B$ is an open ball of radius $\varepsilon$, then the orbit of almost every point enters $B$. To this end, let us call a point *good* if it has a neighborhood such that the orbit of almost every point belonging to it enters $B$. We now wish to show that an arbitrary point $p$ is good. Since $f$ is $\varepsilon$-accessible, there is a *us*-path $[z_0, \dots, z_k]$ with $z_0 \in B$ and $z_k = p$. We shall show by induction on $j$ that each point $z_j$ is good. This is obvious for $j = 0$.

Now suppose that $z_j$ is good. Then $z_j$ has a neighborhood $N$ such that $\mathcal{O}(x) \cap B \neq \emptyset$ for almost every $x \in N$. Let $S$ be the subset of $N$ consisting of points with this property that are also both forward and backward recurrent. Since $f$ preserves a smooth measure it follows from Poincaré's Recurrence Theorem that $S$ has full measure in $N$. If $x \in S$, any point $y \in W^s(x) \cup W^u(x)$ has the property that $\mathcal{O}(y) \cap B \neq \emptyset$. The absolute continuity of the foliations $W^s$ and $W^u$ means that the set

$$\bigcup_{x \in S} (W^s(x) \cup W^u(x))$$

has full measure in the set

$$\bigcup_{x \in N} (W^s(x) \cup W^u(x)).$$

The latter is a neighborhood of $z_{j+1}$. Hence, $z_{j+1}$ is good. □

An immediate consequence of this result is that accessibility implies topological transitivity.

**Theorem 8.3** (Brin, [6]). *Let $f$ be a partially hyperbolic diffeomorphism of a compact Riemannian manifold $M$. Assume that $f$ preserves a smooth measure on $M$ and has the accessibility property. Then for almost every point $x \in M$ the trajectory $\{f^n(x)\}_{n \in \mathbb{Z}}$ is dense in $M$. In particular, $f$ is topologically transitive.*

The assumption that $f$ preserves a smooth measure cannot be dropped in general. Consider the direct product map $F = f \times Id \colon M \times S^1 \to M \times S^1$ where $f$ is a $C^1$ Anosov diffeomorphism of $M$.

**Theorem 8.4** (Niticǎ and Török, [34]). *There exists a $C^1$ open set of $C^1$ small perturbations of $F$ which are accessible but not topologically transitive.*

*Proof.* By Theorem 8.5 below (see also Theorem 8.10) there exists a $C^1$ open and dense set of $C^1$ small perturbations of $F$ which are accessible. Hence, it suffices to construct an open set of non-transitive diffeomorphisms. Choose a map $h \in \mathrm{Diff}^1(S^1)$ as close to the identity as desired such that $h$ has a fixed attracting point. There are two open neghborhoods $U, V \subset S^1$ of this point such that $V$ is not empty, $\bar{V} \subset U$, and $h(\bar{U}) \subset V$. Set $g = f \times h$. We have that $g(M \times \bar{U}) \subset M \times V$ and any map that is $C^0$ close to $g$ has the same property. Note that such a transformation cannot be topologically transitive because each positive semi-orbit has at most one element in the open set $(M \times U) \setminus (M \times V)$. □

**8.3. Stability of accessibility I: $C^1$-genericity.** A diffeomorphism $f$ is said to be *stably accessible* if there exists a neighborhood $\mathfrak{U}$ of $f$ in the space $\mathrm{Diff}^1(M)$ (or in the space $\mathrm{Diff}^1(M, \nu)$ where $\nu$ is an $f$-invariant Borel probability measure) such that any diffeomorphism $g \in \mathfrak{U}$ has the accessibility property.

The study of stable accessibility is based on the *quadrilateral argument* first introduced by Brin (see [7]). This argument is used to establish stable accessibility in various situations. Roughly speaking it goes as follows (we assume for simplicity that the central distribution $E^c$ is integrable). Given a point $p \in M$, consider a 4-legged $us$-path $[z_0, z_1, z_2, z_3, z_4]$ originating at $z_0 = p$. We connect $z_{i-1}$ with $z_i$ by a geodesic $\gamma_i$ lying in the corresponding stable or unstable manifold and we obtain the curve $\Gamma_p = \cup_{1 \le i \le 4} \gamma_i$. We parameterize it by $t \in [0, 1]$ with $\Gamma_p(0) = p$.

If the distribution $E^s \oplus E^u$ were integrable (and hence, the accessibility property for $f$ would fail) the endpoint $z_4 = \Gamma_p(1)$ would lie on the leaf of the corresponding foliation passing through $p$. Therefore, one can hope to achieve accessibility if one can arrange a 4-legged $us$-path in such a way that $\Gamma_p(1) \in W^c(p)$ and $\Gamma_p(1) \neq p$ (see Figure 14). In this case the path $\Gamma_p$ can be homotoped through 4-legged $us$-paths originating at $p$ to the trivial path so that the endpoints stay in $W^c(p)$ during the homotopy and form a continuous curve. Such a situation is usually persistent under small perturbations of $f$ and hence, leads to stable accessibility.

First, we show that the accessibility property is $C^1$ dense in the space of partially hyperbolic (in the narrow sense) diffeomorphisms of $M$, volume preserving or not.

**Theorem 8.5** (Dolgopyat and Wilkinson, [20]). *Let $f \in \mathrm{Diff}^q(M)$ (or $f \in \mathrm{Diff}^q(M, \nu)$, where $\nu$ is a smooth invariant measure on $M$) be partially hyperbolic in the narrow sense. Then for every neighborhood $\mathfrak{U} \subset \mathrm{Diff}^1(M)$*

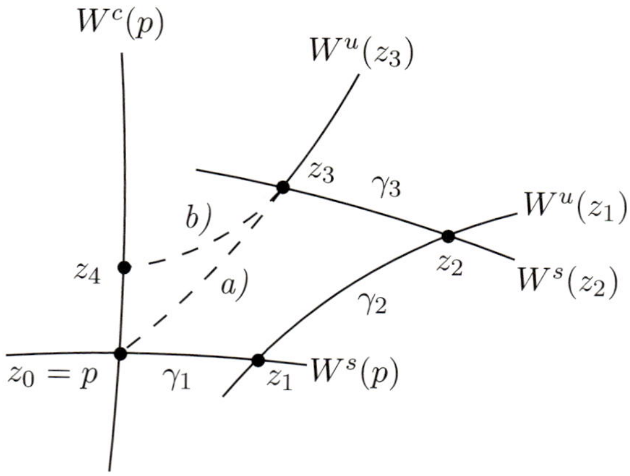


FIGURE 14. Brin's quadrilateral argument: a) $W^s$, $W^u$ are jointly integrable (the curve $\gamma_4$ ends at $z_4 = p$); b) $W^s$, $W^u$ are not jointly integrable (the curve $\gamma_4$ ends at $z_4 \neq p$).

*(respectively,* $\mathfrak{U} \subset \mathrm{Diff}^1(M, \nu)$*) of* $f$ *there exists a* $C^q$ *diffeomorphism* $g \in \mathfrak{U}$ *which is stably accessible.*

*Sketch of the proof.* We shall outline the proof of this theorem in the special case when the central distribution $E^c$ is one-dimensional and integrable. This will allow us to avoid many technical difficulties yet presenting the main idea.

Given $x \in M$ and $r > 0$ denote by $I^c(x, r)$ the interval in $W^c(x)$ centered at $x$ of length $2r$. We call $I^c(x, r)$ the *c-interval* at $x$ (of size $2r$) and we call a finite collection $\mathfrak{I}$ of pairwise disjoint $c$-intervals a *c-family.*

Consider a sufficiently small perturbation $g$ of $f$. We say that $g$ is *accessible modulo a c-family* $\mathfrak{I}$ if there exists $N > 0$ such that the following holds: for every $p, q \in M$ there exist $I_1, \dots, I_n \in \mathfrak{I}$ and a $us$-path for $g$ connecting $p$ with $I_1$, $q$ with $I_n$, and $I_m$ with $I_{m+1}$ ($m = 1, \dots, n-1$) for which the length of each leg and the number of legs do not exceed $N$ (see Figure 15).

If $\mathfrak{I} = \{I^c(x_1, r_1), \dots, I^c(x_n, r_n)\}$ is a $c$-family and $0 < \beta < 1$ we set

$$\beta\mathfrak{I} = \{I^c(x_1, \beta r_1), \dots, I^c(x_n, \beta r_n)\}.$$

We also define $r(\mathfrak{I}) = \max_{1 \le i \le n} r_i$ and $R(\mathfrak{I})$ to be the smallest integer in $\mathbb{N} \cup \{\infty\}$ for which $f^k(I) \cap I' = \emptyset$ for all $I, I' \in \mathfrak{I}$ and $k \le R(\mathfrak{I})$.

Without loss of generality we may assume that periodic points of $f$ are isolated: we can always achieve this by an arbitrarily small perturbation of $f$.

Our first step is to construct a $c$-family with respect to which $f$ is accessible modulo this family. More precisely, the following statement holds.

**Lemma 8.6.** *For every $J > 1$ there exists a c-family $\mathfrak{I}$ such that $r(\mathfrak{I}) < J^{-1}$, $R(\mathfrak{I}) > J$, and $f$ is accessible modulo $0.5\mathfrak{I}$.*

*Proof of the lemma.* By transversality of foliations $W^s$, $W^c$ and $W^u$ such a $c$-family is given by a collection of $c$-intervals $\mathfrak{I} = \{I(x_j, 1)\}$ where the points $x_j$ form an $\varepsilon$-net in $M$ for some sufficiently small $\varepsilon$. If there are some $i, j \le \mathrm{Card}(\mathfrak{I})$ and $k \le J$ such that $f^k(I(x_i, 1)) \cap I(x_j, 1) \neq \emptyset$ we can remove this intersection by an arbitrarily small shift of $x_i$ in the unstable direction. □

We shall now show that accessibility modulo a $c$-family is persistent under small perturbations.

**Lemma 8.7.** *There exists $\delta_0 > 0$ such that for any $0 < \delta < \delta_0$ and any c-family $\mathfrak{I}$ with respect to which $f$ is accessible modulo $0.5\mathfrak{I}$ the following holds: there exists $\sigma > 0$ such that any diffeomorphism $g$ satisfying*

$$d_{C^1}(f, g) < \delta, \quad d_{C^0}(f, g) < \sigma, \tag{8.1}$$

*is accessible modulo $0.75\mathfrak{I}$.*

*Proof of the lemma.* Let $W_1$ and $W_2$ be two foliations of $M$ with smooth leaves (see Section 4.1). We say that $W_1$ is *$\varepsilon$-close* to $W_2$ if given any $p, q$ on the same leaf of $W_1$ with the leafwise distance at most 1 apart, the $W_2$ leaf passing through $p$ intersects the $\varepsilon$ ball centered at $q$.

Let $g$ be a perturbation of $f$ such that $d_{C^1}(f, g) < \delta$. If $\delta_0$ is sufficiently small $g$ is partially hyperbolic. Let $W^s_g$ and $W^u_g$ be its stable and unstable foliations. It is clear that there exists $\varepsilon > 0$ such that $g$ is stably accessible modulo $0.75\mathfrak{I}$ provided that $W^s_g$ is $\varepsilon$-close to $W^s_f$ and $W^u_g$ is $\varepsilon$-close to $W^u_f$. Note that $W^s_g$ and $W^u_g$ depend continuously on $g$ in the $C^1$ topology (see Theorem 3.4). Therefore, $\varepsilon$-closeness would follow if $\delta$ were sufficiently small. However, the number $\delta$ is given and cannot be changed. The meaning of Lemma 8.7 is that $\varepsilon$-closeness of stable and unstable foliations can be achieved provided the maps $f$ and $g$ satisfy (8.1) with $\sigma$ chosen appropriately and independently of $\delta$ (but $\sigma$ depends on $\varepsilon$). We shall prove $\varepsilon$-closeness for stable foliations $W^s_f$ and $W^s_g$. The proof for unstable foliations $W^u_f$ and $W^u_g$ is similar.

We shall show that for all

$$p \in M, \quad q \in B^s(p, 1) \subset W^s_f(p)$$

the intersection $W_g^s(p) \cap B(q, \varepsilon) \neq \emptyset$. Since $f$ is partially hyperbolic in the narrow sense it satisfies (2.7) and (2.8) with contraction and expansion rates satisfying (2.6). There also exists a continuous cone family $C^{cu}(p, \alpha)$ such that for sufficiently small $\alpha > 0$ and any $p \in M$,

(a) $d_p f(C^{cu}(p, \alpha)) \subset C^{cu}(f(p), \alpha)$;
(b) $\|d_p f v\| > \frac{3}{4}\lambda_2 \|v\|$ for any $v \in C^{cu}(p, \alpha)$;
(c) $C^{cu}(p, \alpha)$ is transverse to $E_f^s(p)$ uniformly over $p$.

See (2.11) and (2.12). For sufficiently small $\delta_0 > 0$ the cone $C^{cu}(p, \alpha)$ also satisfies (a)–(c) if $f$ is replaced by any $g$ with $d_{C^1}(f, g) \leq \delta_0$.

Fix $n > 0$ and choose $q \in W^s(p)$ such that $\rho_{W^s}(p, q) \leq 1$. We have that

$$\rho(f^n(p), f^n(q)) \leq \mu_1^n. \tag{8.2}$$

Consider a local smooth submanifold $V$ of dimension $\dim(E^u \oplus E^c)$ passing through $q$ and such that $T_x V \subset C^{cu}(x, \alpha)$ for any $x \in V$ (one can for example, use $V = \exp_q B^{cu}(0, 1)$). By (8.2), for a sufficiently small $\sigma > 0$ the condition $d_{C^0}(f, g) < \sigma$ implies that

$$\rho(g^n(p), g^n(V)) < 2\mu_1^n.$$

By Properties (a)–(c) above, $g^n(V)(y)$ is transverse to $E_g^s(y)$ (as well as $E_f^s(y)$) uniformly over $y \in g^n(V)$. Therefore, there is $K = K(f)$ such that $W_g^s(g^n(p)) \cap g^n(V)$ contains a point $z$ with $\rho(g^n(q), z) \leq K\mu_1^n$. Hence, $g^{-n}(z) \in W_g^s(p)$ and, by Property (b), $\rho(q, g^{-n}(z)) \leq K(\frac{3}{4})^n$. Therefore, if $n$ is large enough the foliation $W_g^s$ is $\varepsilon$-close to $W_f^s$. □

Since (8.1) is an open condition it follows from Lemma 8.7 that any perturbation $g$ satisfying (8.1) is stably accessible modulo $0.75\mathfrak{I}$. Our goal now is to show that there exists an arbitrarily small perturbation $g$ of $f$ which is *accessible on* $0.75\mathfrak{I}$, i.e., for any $c$-interval $I \in \mathfrak{I}$ and any two points $p, q \in 0.75I$ there is a *us*-path connecting $p$ and $q$ (see Figure 15).

**Lemma 8.8.** *Given $\delta > 0$ there exists $J > 1$ with the following property. If $\mathfrak{I}$ is a c-family with $r(\mathfrak{I}) < J^{-1}$ and $R(\mathfrak{I}) > J$ then for any $\sigma > 0$ one can find a diffeomorphism $g$ satisfying* (8.1) *and stably accessible on* $0.75\mathfrak{I}$ *(i.e., any sufficiently small perturbation of $g$ is accessible on* $0.75\mathfrak{I}$*).*

The proof of the theorem now follows from Lemmas 8.6, 8.7 and 8.8. Indeed, given $\delta$, $0 < \delta < \delta_0$ (the number $\delta_0$ is given by Lemma 8.7), choose the number $J$ according to Lemma 8.8. By Lemma 8.6, there exists a $c$-family $\mathfrak{I}$ with $r(\mathfrak{I}) < J^{-1}$ and $R(\mathfrak{I}) > J$ and such that $f$ is accessible modulo $0.5\mathfrak{I}$. Choose $\sigma$ according to Lemma 8.7. Applying Lemma 8.8 we obtain a $C^r$ diffeomorphism $g$ satisfying (8.1) and such that $g$ is stably accessible on

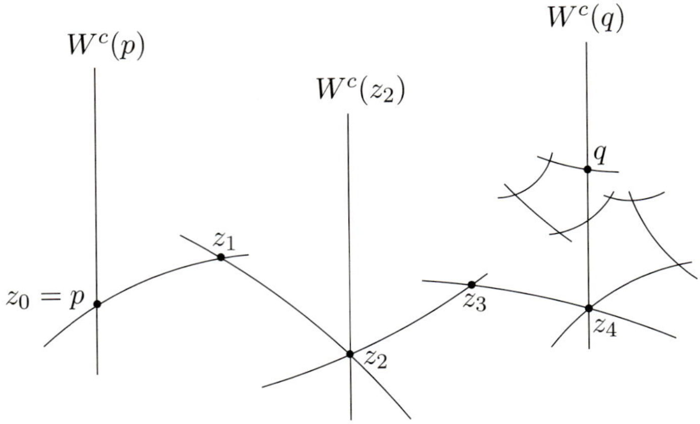


FIGURE 15. To the proof of theorem 8.8: to connect $p$ and $q$ we first connect their central manifolds $W^c(p)$ and $W^c(q)$ by a $us$-path $[z_0, \dots, z_4]$ and then connect $z_4$ with $q$ by a $us$-path "along" $W^c(q)$.

$0.75\mathfrak{J}$. By Lemma 8.7, $g$ is also accessible modulo $0.75\mathfrak{J}$. Thus, $g$ is stably accessible.

We now outline a proof of Lemma 8.8. At this point, for the sake of exposition, we first consider the special case when the distribution $E^s \oplus E^u$ is smooth and integrable and we denote by $W^{su}$ the corresponding foliation. We then show how to modify our arguments to avoid this assumption.

Choose a $c$-interval $I = I(p, r) \in \mathfrak{J}$, a sufficiently small $\varepsilon > 0$, and let

$$U_\varepsilon(I) = \bigcup_{q \in I} B^{su}(q, \varepsilon),$$

where $B^{su}(q, \varepsilon)$ is the ball in $W^{su}(q)$ centered at $q$ of radius $\varepsilon$. The local stable, unstable, and central leaves introduce a local coordinate system in $U_\varepsilon(I)$. The assumption that the distribution $E^s \oplus E^u$ is integrable implies that the holonomy map $\pi$ between different local center-stable manifolds $V^{cs}$ generated by local unstable manifolds $V^u$ preserves the local stable manifolds $V^s$, i.e., for any $q_1, q_2 \in I$,

$$\pi(V^s(q_1)) = V^s(q_2). \tag{8.3}$$

Since $R(\mathfrak{J}) > J$, taking a smaller $\varepsilon$ if necessary we may assume that for any $I, I' \in \mathfrak{J}$ and $0 \leq n \leq J$,

$$f^n(U_\varepsilon(I)) \cap U_\varepsilon(I') = \emptyset. \tag{8.4}$$

Choose any diffeomorphism $\varphi$ of $M$ satisfying

$$\varphi = \mathrm{Id} \text{ outside } U_\varepsilon(I) \text{ and } d_{C^1}(\varphi, \mathrm{Id}) < \delta. \tag{8.5}$$

and consider the perturbation $g = \varphi \circ f$ (so that $d_{C^1}(f, g) < \delta$). We shall estimate the effect of the composition $\varphi \circ f$ on the partially hyperbolic splitting $TM = E^s_g \oplus E^c_g \oplus E^u_g$ for $g$. This is where we make use of the property of the $c$-family $\mathfrak{I}$ that $r(\mathfrak{I}) < J^{-1}$ and $R(\mathfrak{I}) > J$ and our choice of neighborhoods $U_\varepsilon(I)$ (see (8.4)).

**Lemma 8.9.** *There exists $\delta_0 > 0$ such that the following is true. For any $\beta > 0$ there exists $J > 0$ such that for any diffeomorphism $\varphi$ satisfying* (8.5), *and any $p, q \in U_\varepsilon(I)$ with $\rho(p, q) < J^{-1}$ we have*

$$\angle(E^s_g(q), E^s(p)) \le \beta, \quad \angle(d\varphi^{-1}E^u_g(q), E^u(p)) \le \beta.$$

*Proof of the lemma.* By the uniform continuity of the splitting $E^s(p) \oplus E^c(p) \oplus E^u(p)$ in $p$ there exists a positive continuous function $\psi$, with $\psi(0) = 0$, such that for all $p, q \in M$,

$$\begin{aligned} \angle(d\varphi E^u(q), d\varphi E^u(p)) &\le \psi(\rho(p,q)), \\ \angle(E^s(q), E^s(p)) &\le \psi(\rho(p,q)) \end{aligned} \tag{8.6}$$

provided $d_{C^1}(\varphi, \mathrm{Id})$ is sufficiently small. Let

$$\lambda = \max\left\{\frac{\mu_1}{\lambda_2}, \frac{\mu_2}{\lambda_3}\right\} < 1.$$

There exists $C_1 > 0$ and $\theta_0 > 0$ such that for any subspaces $F^s(p), F^u(p) \subset T_pM$ satisfying

$$\max_{p \in M}\{\angle(F^s(p), E^s(p)), \angle(F^u(p), E^u(p))\} \le \theta_0$$

we have that for all $j \ge 0$,

$$\begin{aligned} \angle(df^{-j}F^s(p), df^{-j}E^s(p)) &\le C_1\lambda^j, \\ \angle(df^{j}F^u(p), df^{j}E^u(p)) &\le C_1\lambda^j. \end{aligned}$$

The splitting $E^s_g(p) \oplus E^c_g(p) \oplus E^u_g(p)$ depends continuously on $g$ and therefore,

$$\max_{p \in M}\{\angle(E^s_g(p), E^s(p)), \angle(E^u_g(p), E^u(p))\} \le \theta_0$$

provided $d_{C^1}(\varphi, \mathrm{Id})$ is sufficiently small.

Fix positive $m < J$. If $q \notin U_\varepsilon(I)$ then $g^i(q) = f^i(q)$ for $0 \le i \le m$ and hence,

$$\begin{aligned}\angle(E^s_g(q), E^s(q)) &= \angle(dg^{-m}E^s_g(q), df^{-m}E^s(q))\\ &= \angle(df^{-m}E^s_g(q), df^{-m}E^s(q))\\ &\le C_1\lambda^m.\end{aligned}$$

Similarly, for $q \in U_\varepsilon(I)$ and all $i$ between 1 and $m-1$,

$$g^{-i}(q) = f^{-i}\varphi^{-1}(q) = f^{-i+1}g^{-1}(q)$$

and hence,

$$\begin{aligned}\angle(E^u_g(q), d\varphi E^u(q)) &= \angle(dgE^u_g(q), dgE^u(q))\\ &\le C_2\angle(E^u_g(g^{-1}(q)), E^u(g^{-1}(q)))\\ &\le C_2\angle(df^{m-1}E^u_g(g^{-1}(q)), df^{m-1}E^u(g^{-1}(q)))\\ &\le C_1C_2\lambda^{m-1},\end{aligned}$$

where $C_2 > 0$ is a constant. Combining these inequalities with (8.6) we obtain that

$$\angle(E^s_g(q), E^s(p)) \le C(\lambda^m + \psi(\rho(p,q)))$$

and

$$\angle(d\varphi^{-1}E^u_g(q), E^u(p)) \le 2\angle(E^u_g(q), d\varphi E^u(p)) \le C(\lambda^m + \psi(\rho(p,q))).$$

It follows that if $m$ is sufficiently large and $\rho(p,q)$ is sufficiently small (and hence, $J$ should be sufficiently large) these quantities are bounded by $\beta$. $\square$

Proceeding with the proof of Lemma 8.8 let us pick an arbitrary perturbation $g$ of $f$. Given a point $q \in I$, consider a 4-legged $us$-path $[z_0, z_1, z_2, z_3, z_4]$ for $g$ originating at $z_0 = q$. Without loss of generality we may assume that $z_1 \in W^s_g(z_0)$, $z_2 \in W^u_g(z_1)$, $z_3 \in W^s_g(z_2)$, and $z_4 \in W^u_g(z_3)$. We also assume that the length of each leg in the path does not exceed some small number $r_0$. We connect $z_{i-1}$ with $z_i$ by a geodesic $\gamma_i$ lying in $W^s_g(z_{i-1})$ for $i = 1, 3$ and in $W^u_g(z_{i-1})$ for $i = 2, 4$. Finally, we consider the curve $\Gamma_{g,q} = \cup_{1\le i\le 4}\gamma_i$ and parameterize it by $t \in [0,1]$ with $\Gamma_{g,q}(0) = q$.

By varying the lengths of the last 2 legs in $\Gamma_{g,q}$, one can arrange that $\Gamma_{g,q}(1) \in I$. Furthermore, the path $\Gamma_{g,q}$ can be homotoped through 4-legged $us$-paths originating at $q$ to the trivial path so that the endpoints stay in $I$ during the homotopy. In other words, we obtain a family of 4-legged $us$-paths $\Gamma_{g,q,\alpha}$ originating at $q$ and such that their endpoints $\Gamma_{g,q,\alpha}(1)$ cover an interval $I_1 = I_1(q) \subset I$. Note that $\Gamma_{f,q,\alpha}(1) = q$ for any $q \in I$ and that one may expect this to happen for some perturbations $g$ of $f$.

We now specify a particular choice of the map $\varphi$. Observe that there is a vector field $v$ in $M$ satisfying:

(1) $v$ is divergence free and $\|v\| < \delta$;
(2) $v$ vanishes outside $U_{2\varepsilon}(I)$;
(3) $v(q) \in T_qW^c(q)$ for any $q \in U_\varepsilon(I)$.

This vector field generates a local flow $\varphi_t$ in $U_{2\varepsilon}(I)$. For sufficiently small $t_0 > 0$ the map $\varphi = \varphi_{t_0}$ preserves volume as well as the leaves of $W^c$. It also satisfies

$$d_{C^1}(\varphi, \mathrm{Id}) < \delta, \quad d_{C^0}(\varphi, \mathrm{Id}) < \sigma. \tag{8.7}$$

The map $g_I = \varphi \circ f$ preserves volume and coincides with $f$ outside $U_{2\varepsilon}(I)$. In view of (8.7), it also satisfies (8.1). Using Lemma 8.9 (and choosing $\delta < \delta_0$), the preservation of stable manifolds by the holonomy map $\pi$ (see (8.3)), and Property (3) of the vector field $v$ one can show that for any $q \in 0.75I$ the point $\Gamma_{g,q}(1)$ lies on a positive distance from $q$. Indeed, consider two points moving along $\Gamma_{g,q}$ and $\Gamma_{f,q}$ with the same speed. Observe that they both start at $q$ and the distance between them changes very little while they move along the legs of the paths lying in the stable direction and the distance gains some positive amount while they move along the legs of the paths lying in the unstable direction. We conclude that $g_I$ is accessible and indeed stably accessible on $0.75I$. Composing the maps $g_I$, $I \in \mathfrak{I}$, with each other and with the map $f$ we obtain the desired diffeomorphism $g$.

In the case when the distribution $E^s \oplus E^u$ is neither smooth nor integrable, the main idea is to approximate this distribution locally (within a small neighborhood of each $I \in \mathfrak{I}$) by a smooth and locally integrable distribution and then proceed as above. To this end choose a $c$-interval $I = I(p, r) \in \mathfrak{I}$ and consider a family $\{V(q),\, q \in I\}$ of disjoint smooth local transversals to $I$. For a sufficiently small $\varepsilon > 0$ let

$$U_\varepsilon(I) = \bigcup_{q \in I} B_V(q, \varepsilon),$$

where $B_V(q, \varepsilon)$ is the ball in $V(q)$ centered at $q$ of radius $\varepsilon$. The family of transversals introduces a local coordinate system in $U_\varepsilon(I)$ such that $U_\varepsilon(I) = \xi(D)$, where $D \subset \mathbb{R}^m$ is the unit ball (with $m = \dim M$) and $\xi \colon D \to U_\varepsilon(I)$ a diffeomorphism. In this local coordinates we have a smooth splitting

$$TU_\varepsilon(I) = \widetilde{E}^s \oplus \widetilde{E}^c \oplus \widetilde{E}^u,$$

which is the image under $d\xi$ of a splitting

$$\mathbb{R}^m = \mathbb{R}^s \oplus \mathbb{R}^c \oplus \mathbb{R}^u,$$

($m = \dim M$) and which agrees (by an appropriate choice of $\xi$) with the splitting $E^s \oplus E^c \oplus E^u$ at the point $p$. We also have smooth (local) foliations of $U_\varepsilon(I)$, $\widetilde{V}^s$, $\widetilde{V}^u$, $\widetilde{V}^c$, $\widetilde{V}^{cs}$, $\widetilde{V}^{cu}$, tangent to the corresponding distributions, and such that the foliations $\widetilde{V}^s$ and $\widetilde{V}^u$ are jointly integrable. Note that the distribution $E^c$ and hence, the foliation $\widetilde{V}^c$, is one-dimensional. Finally, we may arrange $\xi$ in such a way that the holonomy map $\widetilde{\pi}$ between different local manifolds $\widetilde{V}^{cs}$ generated by local manifolds $\widetilde{V}^u$ preserves the local manifolds $\widetilde{V}^s$, i.e., for any $q_1, q_2 \in I$,

$$\widetilde{\pi}(\widetilde{V}^s(q_1)) = \widetilde{V}^s(q_2). \tag{8.8}$$

The proof of Lemma 8.9 goes through (with $E^s$ and $E^u$ replaced by $\widetilde{E}^s$ and $\widetilde{E}^u$ respectively) and the construction of the particular map $\varphi$ follows the line of the above argument with some modifications which we now describe.

Fix unit vectors $w^s \in \mathbb{R}^s$ and $w^u \in \mathbb{R}^u$ and define local flows

$$\tau_t^s, \tau_t^u \colon U_\varepsilon(I) \to U_{2\varepsilon}(I)$$

by

$$\tau_t^s(q) = \xi(u + tw^s), \quad \tau_t^u(q) = \xi(u + tw^u), \quad q = \xi u.$$

Note that $\tau_t^s$ sends the $\widetilde{V}^{cs}$ leaves to other $\widetilde{V}^{cs}$ leaves, and between different $\widetilde{V}^{cs}$ leaves it is exactly the holonomy map generated by the $\widetilde{V}^u$ leaves. Similar statement holds for $\tau_t^u$. Moreover, where defined, $\tau_{-t}^s\tau_{-t}^u\tau_t^s\tau_t^u$ is the identity. This expresses the fact that the local foliations $\widetilde{V}^s$ and $\widetilde{V}^u$ are jointly integrable. Finally, there is a vector field $v$ on $M$ satisfying:

(1) $v$ is divergence free and $\|v\| < \delta$;
(2) $v$ vanishes outside $U_{2\varepsilon}(I)$;
(3) $v(q) \in T_q\widetilde{V}^c(q)$ for any $q \in U_\varepsilon(I)$.

This vector field generates a local flow $\varphi_t$ in $U_{2\varepsilon}(I)$. For sufficiently small $t_0 > 0$ the map $\varphi_{t_0}$ satisfies (8.7), preserves volume as well as the leaves of $\widetilde{V}^c$ inside $U_\varepsilon(I)$.

We define the map $\varphi$ as follows

$$\varphi(q) = \tau_{-t}^s\tau_{-t}^u\tau_t^s\tau_t^u\varphi_{t_0}, \quad q \in U_\varepsilon(I)$$

and $\varphi(q) = q$ otherwise. Observe that 1) $\varphi$ preserves volume as well as the leaves of $\widetilde{V}^c$ inside $U_\varepsilon(I)$; 2) $\varphi = \mathrm{Id}$ outside $U_{2\varepsilon}(I)$; 3) $\varphi$ satisfies (8.7).

Similarly to the maps $\tau_t^s$ and $\tau_t^u$ which translate along leaves of $\widetilde{V}^s$ and $\widetilde{V}^u$ respectively, we introduce maps

$$\pi_t^s, \pi_t^u \colon U_\varepsilon(I) \to U_{2\varepsilon}(0.75I)$$

which translate along leaves of $W^s$ and $W^u$:

$$\pi_t^s(q) = W^s(q) \cap \widetilde{V}^{cu}(\tau_t^s(q)), \quad \pi_t^u(q) = W^u(q) \cap \widetilde{V}^{cs}(\tau_t^u(q)).$$

If $\delta$ is sufficiently small these maps are well-defined for sufficiently small $|t|$. Between $\widetilde{V}^{cs}$ leaves $\pi_t^u$ is the holonomy map generated by the leaves of $W^u$ (and similarly for $\pi_t^s$). For sufficiently small $t$ consider the map

$$h = \pi_{-t}^s \pi_t^u \pi_t^s \pi_t^u$$

which is a homeomorphism of $U_\varepsilon(0.75I)$ onto its image. Observe that $h(q)$ is the endpoint of a 4-legged $us$-path for $g$ originating at $q$. The proof now goes to show that

$$\pi_t^u \text{ is approximately } \varphi\tau_t^u \text{ and } \pi_t^s \text{ is approximately } \tau_t^s.$$

This implies that $h$ is approximately $\varphi$. More precisely, one can show that $\rho(h, \varphi) \le t\varepsilon$ for sufficiently small $t$ and that this implies accessibility and indeed, stable accessibility of $g_I = \varphi \circ f$ on $0.75I$. Composing the maps $g_I$, $I \in \mathfrak{I}$, with each other and with the map $f$ we obtain the desired diffeomorphism $g$.

In the general case, when the central foliation is multidimensional and is not necessarily integrable, substantial technical difficulties arise in the construction of the $c$-family in Lemma 8.6 and in the construction of the particular perturbation $\varphi$ in Lemma 8.8. □

**8.4. Stability of accessibility II: particular results.** Theorem 8.5 can be improved in some special cases. We consider skew products over Anosov diffeomorphisms satisfying Condition (2.13), time-$t$ maps of suspension flows and group extensions over Anosov diffeomorphisms. These systems are partially hyperbolic in the narrow sense and so are their small perturbations. Their central distribution is integrable and the corresponding central foliation has compact smooth leaves.

In studying accessibility the following observation is useful. For the sake of definiteness consider a skew product map $F$ over an Anosov diffeomorphism $f\colon M \to M$ where $M$ is compact smooth connected Riemannian manifold. The phase space is fibered with the fiber $N$ where $N$ is a compact smooth Riemannian manifold. Given points $q, p \in M$, there is a $us$-path $[q,p]_f = [z_0, z_1, \dots, z_\ell]_f$ connecting $q$ and $p$. It follows that any point $Q$ in the fiber over $q$ can be connected to some point $P$ in the fiber over $p$ by a $us$-path $[w_0, w_1, \dots, w_\ell]_F$ where each $w_i$ lies in the fiber over $z_i$. Note that the length of each leg in the path and the number of legs are uniformly bounded over $P$ and $Q$. Therefore, if $G$ is a small perturbation of $F$ any point $Q$ can be connected to some point $P$ in the fiber over $p$ by a $us$-path $[Q,P]_G$

(remark that the local stable and unstable manifold depend continuously on the perturbation). If the map $F$ is not accessible we shall show that by an arbitrarily small perturbation one can ensure that any point $P'$ in the fiber over $p$ can be connected to $P$ by a $us$-path $[P, P']_G$. Moreover, any sufficiently small perturbation of $G$ will have the same property.

A similar observation applies to time-$t$ maps of suspension flows and group extensions over Anosov diffeomorphisms (see below). The proofs exploit various versions of Brin's quadrilateral argument (see the beginning of this section).

**1. Skew products over Anosov diffeomorphisms.** We consider the case when the manifold $N$ is one-dimensional, i.e., $N = S^1$. One can strengthen Theorem 8.5 to obtain density of accessibility in the space $\mathrm{Diff}^q(M \times S^1)$ or $\mathrm{Diff}^q(M \times S^1, \nu \times m)$ with $q > 1$ (here $\nu$ is a smooth invariant measure for $f$ and $m$ the length in $S^1$).

**Theorem 8.10** (Nitică and Török, [34]). *Assume that $M$ is a connected manifold. There is a neighborhood $\mathfrak{U}$ of $F$ in $\mathrm{Diff}^q(M \times S^1)$ or $\mathrm{Diff}^q(M \times S^1, \nu \times m)$ such that stable accessibility is open and dense in $\mathfrak{U}$.*

*Sketch of the proof.* We will construct a special small perturbation $G$ of $F$ (in the $C^q$ topology) for which any two points $x$ and $y$ in $W^c_G(p)$, $p \in M$, are accessible and any sufficiently small perturbation of $G$ has the accessibility property. Write $G = F \circ \bar{G}$ where the map $\bar{G}$ coincides with Id outside a small tubular neighborhood of the central leaf $W^c_F(p) = S^1$ and preserves the leaf.

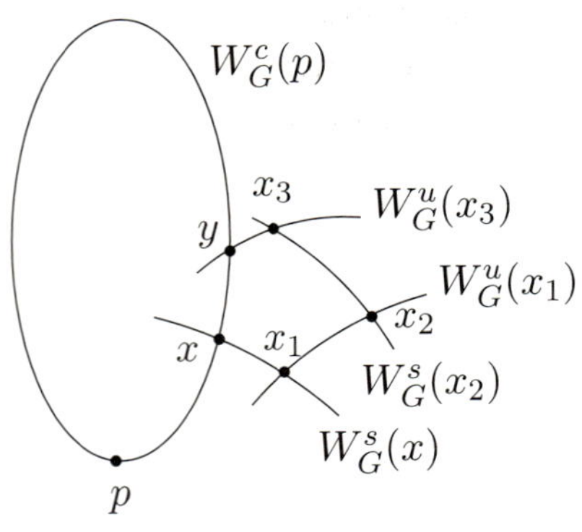


FIGURE 16. To the proof of Theorem 8.13.

Starting from a point $x \in W^c_F(p)\,(= W^c_G(p))$ we construct a point $y \in W^c_F(p)$ in the following way: choose a point $x_1 \in W^s_G(p)$ close to $x$, then a point $x_2 \in W^u_G(x_1)$ close to $x_1$ and finally, a point $x_3 \in W^s_G(x_2)$ close to $x_2$. It is easy to see that the intersection $W^u_G(x_3) \cap W^c_F(p)$ is not empty and consists of a single point (see Figure 16). This is the point $y$. Thus, $y$ is accessible from $x$ and by continuity of the stable and unstable foliations, so is the whole interval between $x$ and $y$. Hence, the following additional conditions ensure that $G$ has the accessibility property:

(1) $\bar{G}|W^c_F(p)$, viewed as a diffeomorphism of $S^1$, has finitely many periodic points $x_1, \dots x_k$;
(2) $y_i \neq x_i$ for all $i$, where $y_i \in W^c_F(p)$ is obtained by the above construction.

Indeed, there is a positive integer $n$ such that

$$\bigcup_{j=-n}^{n} \bigcup_{i=1}^{k} G^j([x_i, y_i]) = W^c_G(p),$$

and therefore, since the intervals $[x_i, y_i]$ are accessible, any two points of $W^c_G(p)$ are accessible. Choose $p$ to be a periodic point of $f$ (and hence, $W^c_F(p)$ to be a periodic central leaf for $F$) of period $m$. We may assume that $m = 1$ (otherwise, replace $F$ by $F^m$ and observe that accessibility of $F^m$ implies accessibility of $F$). One can show that there exists a volume-preserving perturbation $\bar{G}$ which preserves the leaf $W^c_F(p)$ and satisfies the above conditions (1) and (2). Moreover, any perturbation of the diffeomorphism $G$ (in the $C^1$ topology) has the accessibility property. □

**2. Suspension flows.** Let $T_t$ be the suspension flow over a $C^q$ Anosov diffeomorphism $f\colon M \to M$ with a roof function $H\colon M \to \mathbb{R}^+$ and the suspension manifold $\mathcal{M}$ (see Section 2.1).

**Theorem 8.11** (Brin, [9], see also [49]). *There exists an open and dense set $\mathcal{U}$ of $C^q$ functions $H\colon M \to \mathbb{R}^+$ such that the suspension flow $T_t$ is stably accessible.*

*Sketch of the proof.* As discussed above given a point $p \in M$, any point $(q, t) \in \mathcal{M}$, $0 \le t \le H(q)$ is accessible to a point $(p, s)$ for some $0 \le s \le H(p)$. We may assume that $p$ is a periodic point of $f$. Since periodic points are dense, given $\varepsilon > 0$, we can find a periodic point $p_1 \in B(p, \varepsilon)$. Moreover, there is a $us$-path $[x_0, \dots, x_4]$ such that $x_0 = x_4 = p$ and $x_2 = p_1$. Furthermore, given $0 \le t \le H(p)$, there is a unique $us$-path $[z_0, \dots, z_4]$ such that $z_i = (x_i, t_i)$ and $t_0 = t$. If $t_4 \neq t$ the flow $T_t$ is accessible. Indeed, assuming for example, that $\tau = t_4 - t > 0$, by continuity of the stable and unstable foliations we obtain that for any $t \le s \le t_4$ the point $(p, s)$ is accessible from $(p, t)$. Applying the above argument to the map $T_\tau$ we obtain that any point $(p,s)$ with $t_4 \le s \le t_4+\tau$ is accessible from $(p,t_4)$ and so on.

Assume now that $t_4 = t$. We wish to perturb the roof function $H$ such that the new suspension flow $\widetilde{T}_t$ has the accessibility property. Let $\widetilde{H}$ be the roof function of the flow $\widetilde{T}_t$. We may choose $\widetilde{H}$ such that its values along the periodic orbit of $p$ coincide with the corresponding values of $H$ (and hence, the periodic orbit through $p$ for $T_t$ is also a periodic orbit for $\widetilde{T}_t$). Moreover, we may choose $\widetilde{H}$ such that the above procedure, applied to the flow $\widetilde{T}_t$, will result in $t_4 \ne t$.

Let $n$ be the period of $p$ (as a periodic point of $f$). Choose $\varepsilon_0 > 0$ so small that

(p1) $x_1, x_2, x_3 \notin B(p,\varepsilon_0)$;
(p2) $B(p,\varepsilon_0) \cap f^i(B(p,\varepsilon_0)) = \emptyset$ for $i = 1,\dots,n-1$;
(p3) $B(p,\varepsilon_0) \cap V^s_f(p_1) = \emptyset$ and $B(p,\varepsilon_0) \cap V^u_f(p_1) = \emptyset$.

Choose $\varepsilon_1 > \varepsilon_0$ such that $x_1 \notin B(p,\varepsilon_1)$. Given $\delta > 0$, set $\widetilde{H} = H - h(\rho(p,x))$ where $h$ is a $C^\infty$ function satisfying:

(h1) $h(u) = 0$ if $u \ge \varepsilon_1$;
(h2) $h(u) > 0$ if $u < \varepsilon_1$;
(h3) $h(u) = \delta/2$ if $u \le \varepsilon_0$;
(h4) $\|h(u)\|_{C^r} \le \delta$ for any $1 \le r \le q$.

By construction, the flows $\widetilde{T}_t$ coincides with the flow $T_t$ in a neighborhood of the periodic orbit through $p_1$. It follows that these flows have the same local stable and unstable manifolds at $(p_1,t)$ for any $t$. Moreover, there is a unique $us$-path $[\widetilde{z}_0,\dots,\widetilde{z}_4]$ such that $\widetilde{z}_i = (x_i,\widetilde{t}_i)$ and $\widetilde{t}_0 = t$. By (p1) – (p3) and (h1) – (h4), we obtain that

$$\widetilde{t}_1 > t_1 + \delta/2, \quad \widetilde{t}_4 > t_4 + \delta/2$$

and

$$\widetilde{t}_1 - t_1 = \widetilde{t}_2 - t_2 = \widetilde{t}_3 - t_3.$$

The desired result follows. □

**3. Group extensions.** Let $G$ be a compact connected Lie group, $f\colon M \to M$ a $C^q$ Anosov diffeomorphism, and $\varphi\colon M \to G$ a $C^q$ function. Consider the $G$-extension $F = F_\varphi\colon M \times G \to M \times G$ of $f$,

$$F_\varphi(x,y) = (f(x), \varphi(x)y). \tag{8.9}$$

See Section 2.3.

**Theorem 8.12** (Brin, [7], Burns and Wilkinson, [17]). *For every neighborhood $\mathfrak{U} \subset C^q(M,G)$ of the function $\varphi$ there exists a function $\psi \in \mathfrak{U}$ such that the* *diffeomorphism $F_\psi$ is stably accessible. In other words, stably accessible group extensions are dense in the space of $C^q$ group extensions over the Anosov diffeomorphism $f$.*

We outline a proof of this theorem following the approach by Burns and Wilkinson [17] which is built upon some earlier ideas from the work of Brin [7]. We say that the two points $(x_1, g_1), (x_2, g_2) \in M \times G$ are *equivalent* and write $(x_1, g_1) \sim (x_2, g_2)$ if there is a *us*-path from $(x_1, g_1)$ to $(x_2, g_2)$. We also say that the points $(x, g_1), (x, g_2) \in M \times G$ are *restricted equivalent* and write $(x, g_1) \overset{\circ}{\sim} (x, g_2)$ if there is a *us*-path from $(x, g_1)$ to $(x, g_2)$ whose projection to $M$ is a null-homotopic loop. It is easy to see that for every $g \in G$ these relations are invariant under the right translation $R_g$ of $G \times M$ (that sends $(x, h)$ to $(x, hg)$) and that each equivalence class of the relation $\sim$ meets $\{x\} \times G$ for any $x \in M$ (since any point $(y, g)$ can be connected to a point on the fiber $\{x\} \times G$).

The *holonomy group* for $F_\varphi$ at $x_0$ is defined by

$$H = H(x_0, F_\varphi) = \{g \in G \colon (x_0, g) \sim (x_0, e)\}.$$

We also define the *restricted holonomy group* for $F_\varphi$ at $x_0$ by

$$H^0 = H^0(x_0, F_\varphi) = \{g \in G \colon (x_0, g) \overset{\circ}{\sim} (x_0, e)\}.$$

It is easy to see that $H$ is a Lie subgroup of $G$. So is $H^0$. Indeed, it is an analytic subgroup (i.e., a connected Lie subgroup) of $G$. To see this note that for any element $g \in H^0$, there is a *us*-path $\gamma : [0, 1] \to M \times G$ such that $\gamma(0) = (x_0, e)$, $\gamma(1) = (x_0, g)$, and $\pi_M \circ \gamma$ ($\pi_M$ is the natural projection to $M$) is homotopic to the trivial loop through a homotopy fixing the basepoint $x_0$. One can show that there is a homotopy through *us*-loops in $M$ (for the map $f$) based at $x_0$ from the trivial loop to $\pi_M \circ \gamma$. One can further show that each path in this homotopy has a lift to a *us*-path in $M \times G$ that starts at $(x_0, e)$ and ends in $\pi_M^{-1}(x_0)$. This gives us a continuous map $A : [0, 1] \times [0, 1] \to M \times G$ such that, for each $t \in [0, 1]$, the path $A(t, \cdot)$ is a *us*-path in $M \times G$ starting at $(x_0, e)$, covering a null-homotopic loop in $M$ and ending in the fiber over $x_0$. Moreover $A(0, \cdot)$ is the trivial loop at $(x_0, e)$ and $A(1, 1) = \gamma(1) = (x_0, g)$. The path $\pi_G \circ A(\cdot, 1)$ ($\pi_G$ is the natural projection to $G$) is a path in $H^0$ from $e$ to $g$. Thus, $H^0$ is path-connected. By a theorem of Kuranashi-Yamabe [50], a path-connected subgroup of a Lie group is an analytic Lie subgroup.

The desired result is an immediate corollary of the following two statements.

**Lemma 8.13.** *If $H^0 = H^0(F_\varphi) = G$, then $F_\varphi$ is stably accessible.*

**Lemma 8.14.** *For any $\varepsilon > 0$ there exists a $C^q$ group extension $F_\psi$ such that $d_{C^q}(F_\varphi, F_\psi) \le \varepsilon$ and $H^0(F_\psi) = G$.*

We outline the proof of Lemma 8.13. Since $H^0 = G$, every $x \in M$ has the property that every point in $\{x\} \times G$ can be reached from $(x, e)$ along a $us$-path. Since every $x \in M$ can be reached from $x_0$ along a $us$-path (for the map $f$) it follows that every $(x, g) \in M \times G$ can be reached from $(x_0, e)$ along a $us$-path. It suffices to show that $(x_0, e)$ can be connected by a $us$-path to a point $(x_0, g)$ with $g \ne e$.

Let $Z$ be a set. We call a map $h\colon Z \to G$ *achievable* if it is the "endpoint map" of a continuous family of $us$-paths that begin at $(x_0, e)$, end in $\{x_0\} \times G$ and project to null-homotopic loops in $M$. More precisely, $h$ is achievable if there are a continuous function $H\colon Z \times [0,1] \to M \times G$ and a positive integer $C$ such that for each $z \in Z$, the path $h_z\colon [0,1] \to M \times G$ defined by $h_z(t) = H(z,t)$ is a $us$-path from $(x_0, e)$ to $(x_0, h(z))$ with at most $C$ legs that is the lift of a null-homotopic loop in $M$.

It suffices to construct an achievable map $h\colon (D, \partial D) \to (G, G \setminus \{e\})$ (where $D$ is homeomorphic to the closed disc of dimension $n = \dim G$) that induces a nontrivial map on $n$-dimensional homology. Moreover, it is not necessary that $h$ itself be achievable but it can be approximated arbitrarily closely in the $C^0$ topology by an achievable map.

To this end we call a map $h : Z \to G$ *approximable* if for each $\varepsilon > 0$ there is an achievable map $h^\varepsilon\colon Z \to G$ such that $\mathrm{dist}_{C^0}(h, h^\varepsilon) < \varepsilon$.

One can show that if $h_i : Z_i \to G$, $1 \le i \le k$ are approximable then the product map

$$(z_1, \dots, z_k) \mapsto h_k(z_k) h_{k-1}(z_{k-1}) \cdots h_2(z_2) h_1(z_1)$$

is approximable.

Suppose now that we knew that the geodesic arc $\sigma_v : [0,1] \to G$, $\sigma_v(t) = \exp(tv)$, was approximable for all $v$ in the Lie algebra $\mathfrak{L}(G)$. Then the map

$$h(z_1, \dots, z_n) = \exp(z_n v_n) \exp(z_{n-1} v_{n-1}) \cdots \exp(z_2 v_2) \exp(z_1 v_1)$$

is approximable for any choice of $v_1, \dots, v_n \in \mathfrak{L}(G)$. Moreover, $h$ will be a diffeomorphism of $D = [-1,1]^n$ onto a neighborhood of $e$ if we choose the vectors $v_i$ to be short enough and to form a basis for $\mathfrak{L}(G)$ that is orthogonal with respect to the bi-invariant metric.

Let us call $v \in \mathfrak{L}(G)$ *approximable* if $\sigma_v$ is approximable. The proof goes on by showing that the set of approximable vectors is a Lie subalgebra of $\mathfrak{L}(G)$, i.e., if a vector $v \in \mathfrak{L}(G)$ is approximable then so is the vector $\lambda v$ for all $\lambda \in \mathbb{R}$ and if vectors $v$ and $w$ are approximable then so are the vectors

$v + w$ and $[v, w]$ (the Lie bracket of $v$ and $w$). The hard part of the proof is to show that this subalgebra coincides with $H^0$, in other words, the geodesic arc $\sigma_v$ is approximable if and only if $v$ is tangent to $H^0$. The lemma follows.

We now sketch a proof of Lemma 8.14. First, we show that $\varphi$ can be perturbed so as to make $H = G$. Since periodic points for $f$ are dense in $M$ we may choose the base-point $x_0$ of our holonomy groups to be a periodic point. Since both $W^u(x_0)$ and $W^s(x_0)$ are dense in $M$ there are infinitely many distinct homoclinic orbits for $x_0$. Let $z_1, \dots, z_n$, $n = \dim G$, be homoclinic points for $x_0$ that belong to distinct orbits of $f$. For each $z_i$ choose a neighborhood $U_i$ such that $x_0 \notin U_i$ and $f^k(z_j) \in U_i$ if and only if $j = i$ and $k = 0$. Now choose for each $i$ a two-legged $us$-loop formed by a path from $x_0$ to $z_i$ in $W^u(x_0)$ and a path from $z_i$ to $x_0$ in $W^s(x_0)$. Let $\gamma_i$, $1 \le i \le n$, be the lifts of these loops to $us$-paths that begin at $(x_0, e)$ and end in $\{x_0\} \times G$. If $\widetilde{\varphi}$ is a perturbation of $\varphi$, let $\widetilde{\gamma}_i$ be the $us$-path for $F_{\widetilde{\varphi}}$ that starts at $(x_0, e)$ and has the same projection to $M$ as $\gamma_i$. Observe that if $\varphi$ and $\widetilde{\varphi}$ differ only in $U_i$, then $\gamma_j$ and $\widetilde{\gamma}_j$ both have the same endpoint for each $j \ne i$. It is not difficult to perturb $\varphi$ within $U_i$ so as to move the endpoint of $\widetilde{\gamma}_i$ in any desired direction within $\{x_0\} \times G$. Consequently one can arrange that the projections to $G$ of the endpoints of the paths $\widetilde{\gamma}_1, \dots, \widetilde{\gamma}_n$ do not lie in any subgroup of $G$ with dimension less than $n$. Since these projections belong to $H(f_{\widetilde{\varphi}})$, we obtain $H(f_{\widetilde{\varphi}}) = G$.

In order to adapt the preceding argument so that it applies to $H^0$, we need to change the paths $\gamma_i$ so that their projections to $M$ are null homotopic. To this end, we choose periodic points $x_i$, $1 \le i \le n$, close to $x_0$, and for each $i$ a pair of heteroclinic points $z_i^0$ and $z_i^1$ such that there are short paths from $x_0$ to $z_i^0$ in $W^u(x_0)$, from $z_i^0$ to $x_i$ in $W^s(x_i)$, from $x_i$ to $z_i^1$ in $W^u(x_i)$, and from $z_i^1$ to $x_0$ in $W^s(x_0)$. This loop is null homotopic and its lift to a $us$-path starting at $(x_0, e)$ is the new $\gamma_i$. Of course, we ensure that all of the points were chosen from different orbits of $f$. The sets $U_i$ are now neighborhoods of the points $z_i^0$ with the properties that none of the periodic points $x_0, x_1, \dots, x_n$ is in any $U_i$ and $f^k(z_j^a) \in U_i$ if and only if $j = i$, $a = 0$ and $k = 0$. After these changes, the argument proceeds as before and the lemma follows.

This completes the proof of Theorem 8.12.

**4. Time-$t$ maps of an Anosov flow.** Let $\varphi_t$ be an Anosov flow on a compact smooth Riemannian manifold $M$. It turns out that stable accessibility of the time-1 diffeomorphism depends on whether the distribution $E^s \oplus E^u$ is integrable, i.e., whether the stable and unstable foliations, $W^s$ and $W^u$, of the time-1 map are jointly integrable. First, let us comment on joint integrability.

Fix $\varepsilon > 0$. Given a point $x \in M$, consider a local smooth submanifold through $x$,

$$\Pi(x) = \bigcup_{y \in B^u(x,\varepsilon)} \bigcup_{-\varepsilon \le \tau \le \varepsilon} \varphi_\tau(y).$$

For $x, x' \in M$ let $\pi_{x,x'} \colon \Pi(x) \to \Pi(x')$ be the holonomy map generated by the family of local stable manifolds. The foliations $W^s$ and $W^u$ are *jointly integrable* if for every $y \in \Pi(x)$ the image of the local unstable leaf $V^u(y)$ under $\pi_{x,x'}$ is the local unstable leaf $V^u(\pi_{x,x'}(y))$.

**Theorem 8.15** (Burns, Pugh and Wilkinson, [16]). *Assume that the stable and unstable foliations of the flow are not jointly integrable. Then the time-1 map $\varphi_1$ is stably accessible.*

*Sketch of the proof.* By a result of Plante [38] if the stable and unstable foliations of the flow are not jointly integrable then all leaves of both $W^s$ and $W^u$ are dense in $M$. Given $\varepsilon > 0$ there exists $\delta > 0$ such that any diffeomorphism $g$ which is $\delta$-close to $\varphi_1$ in the $C^1$ topology has the property that every leaf of either $W^s$ or $W^u$ is $\varepsilon$-dense in $M$ (i.e., for any $x, y \in M$ either $W^s(x) \cap B(y, \varepsilon)$ or $W^u(x) \cap B(y, \varepsilon)$ are not empty). It follows that for any $x, y \in M$ there is a point $w \in B(y, \varepsilon)$ which is accessible from $x$. Therefore, one can find $z = \varphi_\tau(y)$ with $|\tau| \le 2\varepsilon$ which is accessible from $x$.

Note that if the foliations $W^s$ and $W^u$ are *not* jointly integrable then there is a point $y \in M$ such that the following property holds: one can find a point $y' = \varphi_{\tau'}(y)$ with $\tau' \neq 0$ and a four-legged $us$-path $[y, y']_{\varphi_1}$ starting at $y$ and ending at $y'$. One can then show that for a sufficiently small $\varepsilon$ there is a point $y'' = \varphi_{\tau''}(y)$ with $\tau'' \neq 0$ and a four-legged $us$-path $[y, y'']_g$ starting at $y$ and ending at $y''$. It follows that any point between $y$ and $y''$ is accessible from $y$. Applying the map $\varphi_{\tau''}$ to the point $y$ we obtain that any point between $\varphi_{\tau''}(y) = y''$ and $\varphi_{\tau''}(y'')$ is accessible from $y''$ and so on. Hence, the point $z$ is accessible from $y$ and the desired result follows. □

By verifying conditions of Theorem 8.15 one can establish stable accessibility of the time-1 map for the following flows:

(1) geodesic flows on negatively curved manifolds (more generally, contact flows; Katok and Kononenko, [30]);
(2) $C^2$ volume-preserving flows on compact 3-manifolds which are not special flows with a constant height function (Burns, Pugh and Wilkinson, [16]).

# 9. The Pugh–Shub Ergodicity Theory

Let $f$ be a $C^2$ diffeomorphism of a smooth compact Riemannian manifold $M$ which is partially hyperbolic in the narrow sense. Assume that $f$ preserves a smooth measure $\nu$. Let $TM = E^s \oplus E^c \oplus E^u$ be the invariant splitting of the tangent bundle such that

$$\lambda < \|df|E^s\| < \mu, \quad \gamma < \|df|E^c\| < \gamma^{-1}, \quad \mu^{-1} < \|df|E^u\| < \lambda^{-1},$$

where $0 < \lambda < \mu < \gamma \le 1$.

To study ergodicity of $f$ we use a version of the Hopf argument that we have exploited to prove ergodicity of Anosov diffeomorphisms (see the proof of Theorem 7.9). This version is better adapted to the case of partially hyperbolic systems.

Let $\mathcal{B}$ be the Borel $\sigma$-algebra of $M$. Call $x, y \in M$ stably equivalent if

$$\rho(f^n(x), f^n(y)) \to 0 \text{ as } n \to +\infty,$$

and unstably equivalent if

$$\rho(f^n(x), f^n(y)) \to 0 \text{ as } n \to -\infty.$$

Stable and unstable equivalence classes induce two partitions of $M$, and we denote by $\mathcal{S}$, respectively, $\mathcal{U}$, the Borel $\sigma$-algebra they generate. Recall that for an algebra $\mathcal{A} \subset \mathcal{B}$ its saturated algebra is the set

$$Sat(\mathcal{A}) = \{B \in \mathcal{B} : \text{ there exists } A \in \mathcal{A} \text{ with } \nu(A \triangle B) = 0\}.$$

It follows from the Hopf decomposition theorem that $f$ is ergodic if

$$Sat(\mathcal{S}) \cap Sat(\mathcal{U}) = \mathcal{T}, \tag{9.1}$$

where $\mathcal{T}$ is the trivial algebra.

For an Anosov diffeomorphism $f$ the stable equivalent class containing a point $x$ is the leaf $W^s(x)$ of the stable foliation. Similarly, the unstable equivalent class containing $x$ is the leaf $W^u(x)$ of the unstable foliation. The $\sigma$-algebra $\mathcal{S}$ consists of those Borel sets $S$ for which $W^s(x) \subset S$ whenever $x \in S$. Respectively, the $\sigma$-algebra $\mathcal{U}$ consists of those Borel sets $U$ for which $W^u(x) \subset U$ whenever $x \in U$. The relation (9.1) holds due to the absolute continuity of stable and unstable foliations thus, proving ergodicity for Anosov diffeomorphisms.

If a diffeomorphism $f$ is partially hyperbolic in the narrow sense the stable and unstable foliations $W^s$ and $W^u$ of $M$ also generate Borel $\sigma$-algebras $\mathcal{M}^s$ and $\mathcal{M}^u$, respectively, and we have that $\mathcal{S} \subset \mathcal{M}^s$ and $\mathcal{U} \subset \mathcal{M}^u$ (note that stable and unstable sets containing a point $x$ may be larger than $W^s(x)$

and $W^u(x)$ due to possible contractions and expansions along the central directions). It follows that

$$Sat(\mathcal{S}) \cap Sat(\mathcal{U}) \subset Sat(\mathcal{M}^s) \cap Sat(\mathcal{M}^u).$$

If the map $f$ is accessible we have that $Sat(\mathcal{M}^s \cap \mathcal{M}^u) = \mathcal{T}$. In fact one can require a weaker property than accessibility to make the same conclusion. We say that $f$ possesses the *essential accessibility property* if the partition of $M$ by the accessibility classes is trivial in the measure-theoretical sense, i.e., any measurable set which consists of accessibility classes has measure zero or one.

If $f$ is essentially accessible then ergodicity would follow if we could establish that

$$Sat(\mathcal{M}^s) \cap Sat(\mathcal{M}^u) \subset Sat(\mathcal{M}^s \cap \mathcal{M}^u) \tag{9.2}$$

(note that the opposite inclusion is obvious). We shall describe some conditions which guarantee (9.2).

Recall that the center-stable foliation $W^{cs}$ is absolutely continuous if the holonomy map generated by the family of local leaves $V^{cs}$ of the foliation is absolutely continuous (see Section 7.1). In other words the conditional measure generated by the volume on the local leaves is absolutely continuous with respect to the volume on the leaves. A similar notion applies to the center-unstable foliation $W^{cu}$.

**Theorem 9.1.** *Let $f$ be a volume-preserving partially hyperbolic diffeomorphism. Assume that that $f$ is essentially accessible and dynamically coherent. Assume also that the foliations $W^{cs}$ and $W^{cu}$ are absolutely continuous. Then $f$ is ergodic.*

*Proof.* Inclusion 9.2 is an immediate corollary of the conditions of the theorem. □

The assumption that the foliations $W^{cs}$ and $W^{cu}$ are absolutely continuous is very strong. It holds for example, when the center foliation is smooth (more generally, Lipschitz continuous). However, "typically" the center foliation is not even absolutely continuous (see Section 10.3). We will use more "practical" yet still technical assumption.

We say that $f$ is *center bunched* if

$$\mu^\theta < \gamma, \text{ where } \theta = \frac{\log \mu - \log \gamma}{\log \lambda}$$

(i.e., $\mu = \gamma\lambda^\theta$). The number $\theta$ is the Hölder exponent of the distributions $E^s$, $E^c$, and $E^u$ (compare with (3.2)).

**Theorem 9.2** (Grayson, Pugh and Shub, [21], see also [41]). *Assume that a partially hyperbolic diffeomorphism $f$ is (essentially) accessible, dynamically coherent, and center bunched. Then $f$ is ergodic.*

**Remark 9.3.** It is an open problem whether the theorem holds without the technical hypothesis that $f$ is dynamically coherent and center bunched.

*Sketch of the proof of the theorem.* Our goal is to establish (9.2). To this end given a point $x \in M$, we introduce a collection of special sets at $x$ called *Juliennes* , $J_n(x)$ (they resemble slivered vegetables). We shall describe a construction of these sets which assures that

(J1) $J_n(x)$ form a basis of the topology.

(J2) $J_n(x)$ form a basis of the Borel $\sigma$-algebra. More precisely, let $Z$ be a Borel set; a point $x \in Z$ is called *Julienne dense* if
$$\lim_{n\to+\infty} \frac{\nu(J_n(x)\cap Z)}{\nu(J_n(x))} = 1.$$
Let $D(Z)$ be the set of all Julienne dense points of $Z$. Then
$$D(Z) = Z \pmod 0.$$

(J3) If $Z \in Sat(\mathcal{M}^s) \cap Sat(\mathcal{M}^u)$, then $\mathcal{D}(Z) \in Sat(\mathcal{M}^s \cap \mathcal{M}^u)$.

Properties (J1)–(J3) imply (9.2).

Note that the collection of balls $B(x, \frac{1}{n})$ satisfies requirements (J1) and (J2) but not (J3). Juliennes can be viewed as balls "distorted" by the dynamics in the following sense. Fix an integer $n \geq 0$, a point $x \in M$ and numbers $\tau, \sigma$ such that $0 < \tau < \sigma < 1$. Denote by
$$B^s_n(x,\tau) = \{y \in W^s(x)\colon \rho(f^{-k}(x), f^{-k}(y)) \leq \tau^k\},$$
$$B^u_n(x,\tau) = \{y \in W^u(x)\colon \rho(f^{k}(x), f^{k}(y)) \leq \tau^k\}.$$
The Julienne $J_n(x)$ is defined as the local foliation product
$$J_n(x) = B^s_n(x,\tau) \times B^u_n(x,\tau) \times B^c(x,\sigma^n),$$
where $B^c(x,\sigma^n)$ is the ball in $W^c(x)$ centered at $x$ of radius $\sigma^n$. We also introduce *center stable* and *center unstable juliennes* as the local foliation product
$$J^{cs}_n(x) = B^s_n(x,\tau) \times B^c(x,\sigma^n), \quad J^{cu}_n(x) = B^u_n(x,\tau) \times B^c(x,\sigma^n).$$
The proof of $(J1)-(J3)$ is based on the the following properties of juliennes:

(1) *scaling*: for any fixed $k \geq 0$,
$$\frac{\nu(J_n(x))}{\nu(J_{n+k}(x))}$$
is uniformly bounded as $n \to \infty$;

(2) *engulfing*: there is $\ell \geq 0$ such that, for any $x, y \in M$, if the intersection $J_{n+\ell}(x) \cap J_{n+\ell}(y)$ is not empty then

$$J_{n+\ell}(x) \cup J_{n+\ell}(y) \subset J_n(x);$$

(3) *quasi-conformality*: there is $k \geq 0$ such that if $x, y \in M$ are connected by an arc on an unstable manifold that has length $\leq 1$ then the holonomy map $\pi : V^{cs}(x) \to V^{cs}(y)$ generated by the family of local unstable manifolds (see Section 7.1) satisfies

$$J^{cs}_{n+k}(y) \subset \pi(J^{cs}_n(x)) \subset J^{cs}_{n-k}(y).$$

The properties (1) and (2) are possessed by the family of balls in Euclidean space and they underline the proof of the Lebesgue Density Theorem. One can use these properties to show that juliennes are density bases. The center-unstable juliennes are a density basis on $W^{cu}(x)$ with respect to the smooth conditional measure $\nu_{W^{cu}}$ on $W^{cu}(x)$, the center-stable juliennes are a density basis on $W^{cs}(x)$ with respect to the smooth conditional measure $\nu_{W^{cs}}$ on $W^{cs}(x)$, and the juliennes are a density basis on $M$ with respect to the smooth measure $\nu$.

Juliennes, $J_n(x)$, are small but highly eccentric sets in the sense that the ratio of their diameter to their inner diameter increases with $n$ (the inner diameter of a set is the diameter of the largest ball it contains). In general, sets of such shape may not form density bases as an example by Nikodym [33] of a "paradoxical set" $N$ in the plane shows. This set is contained in the unit square in plane, it has full Lebesgue measure there, and each $p \in N$ lies on a line meeting $N$ only at $p$. Therefore, $N$ has a closed subset $K$ of positive measure with the same property. For each $p \in K$ draw a line $L$ through $p$ that meets $K$ only at $p$. Draw a rectangle $R$ along $L$, centered at $p$, such that its side parallel to $L$ has length $\epsilon$, and its side perpendicular to $L$ has length $\delta$. Fix some $\mu$, $0 < \mu < 1$. When $\delta \ll \mu\epsilon$, $R \cap K$ is contained in the $(\mu\epsilon, \delta)$-part of $R$ at $p$. (Otherwise, by closedness of $K$ there would be additional points of $K$ on $L$.) Hence,

$$\frac{m(R \cap K)}{m(R)} \leq \frac{\mu\epsilon\delta}{\epsilon\delta} = \mu,$$

where $m$ is area. Since $\mu$ is arbitrarily small, there is a sequence of rectangles $R_n$ shrinking down to $p$ such that

$$\lim_{n\to\infty} \frac{m(R_n \cap K)}{m(R_n)} = 0.$$

Thus, a set such as $K$ has positive measure but each point $p$ has density zero with respect to rectangles that shrink badly to $p$.[5]

What makes juliennes different from Nikodym's rectangles is that their elongation and eccentricity are controlled by the dynamics, in particular, they nest in a way similar to balls.

Quasi-conformality is what is needed to prove Property (J3). Roughly speaking it means that the holonomy map (almost) preserves the shape of juliennes. □

---

[5]I would like to thank C. Pugh for providing me with this description of the Nikodym set.

## 10. Stable Ergodicity

**10.1. The Pugh–Shub Stable Ergodicity Theorem.** A diffeomorphism $f$ of a smooth compact Riemannian manifold $M$ preserving a smooth measure $\nu$ is called *stably ergodic* if any small perturbation of $f$ preserving $\nu$ is ergodic.

To establish stable ergodicity of a partially hyperbolic diffeomorphism $f$ let us analyze whether conditions of Theorem 9.2 are stable under small perturbations:

(a) if $f$ is dynamically coherent and the central foliation $W^c$ is of class $C^1$ then every diffeomorphism $g$ which is sufficiently close to $f$ in the $C^1$ topology is dynamically coherent (see Theorems 5.10 and 5.14);

(b) if $f$ is center bunched then every diffeomorphism $g$ which is sufficiently close to $f$ in the $C^1$ topology is center bunched.

This shows that for partially hyperbolic diffeomorphisms which are dynamically coherent and center bunched stable ergodicity amounts to stable accessibility. More precisely, the following theorem holds.

**Theorem 10.1.** *Assume that a partially hyperbolic diffeomorphism $f$ preserves a smooth measure $\nu$, is dynamically coherent and center bunched. If $f$ is (essentially) stably accessible, then $f$ is stably ergodic.*

Combining this theorem with results in Section 8.4 we obtain the following classes of stably ergodic systems:

1. *skew product maps over Anosov diffeomorphisms*: if $F = f\times\mathrm{Id} : M\times S^1 \to M\times S^1$ then there is a neighborhood $\mathcal{U}$ of $F$ in $\mathrm{Diff}^2(M\times S^1)$ or $\mathrm{Diff}^2(M\times S^1, \nu\times m)$ such that stable ergodicity is open and dense in $\mathcal{U}$ (here $m$ is the length);
2. *suspension flows over Anosov diffeomorphisms*: there exists an open and dense set of $C^q$ functions $H\colon M\to\mathbb{R}^+$ such that the suspension flow $T_t$ with the roof function $H$ is stably ergodic;
3. *group extensions over Anosov diffeomorphisms*: if $F_\varphi\colon M\times G\to M\times G$ is a group extension then for every neighborhood $\mathcal{U}\subset C^q(M,G)$ of the function $\varphi$ there exists a function $\psi\in\mathcal{U}$ such that the diffeomorphism $F_\psi$ is stably ergodic;
4. *time-$t$ maps of Anosov flows*: if the stable and unstable foliations of an Anosov flow are not jointly integrable then the time-$t$ map for $t\neq 0$ is stably ergodic.

**10.2. Frame flows.** We describe some cases in which the frame flow is ergodic.

**Theorem 10.2.** *Let $\Phi_t$ be the frame flow on a compact smooth Riemannian $n$-dimensional manifold $M$ of negative sectional curvature pinched between $-\Lambda^2$ and $-\lambda^2$. Then the flow is ergodic in the following cases:*

(1) *if $M$ has constant negative curvature (Brin,* [6]*);*
(2) *for a set of metrics of negative curvature which is open and dense in the $C^3$ topology (Brin,* [6]*);*
(3) *if $n$ is odd and $n \neq 7$ (Brin and Gromov,* [11]*);*
(4) *if $n$ is even, $n \neq 8$, and $\lambda/\Lambda > 0.93$ (Brin and Karcher,* [12]*)*
(5) *if $n = 7$ and $\lambda/\Lambda > 0.999714...$ (Burns and Pollicott,* [15]*);*
(6) *if $n = 8$ and $\lambda/\Lambda > 0.999785...$ (Burns and Pollicott,* [15]*).*

**10.3. Pathological foliations.** Let $A$ be an area-preserving linear hyperbolic automorphism of the two dimensional torus $\mathbb{T}^2$. Consider the map $F = A \times \mathrm{Id}$ of the three dimensional torus $\mathbb{T}^3 = \mathbb{T}^2 \times S^1$. Any sufficiently small $C^1$ perturbation $G$ of $F$ is uniformly partially hyperbolic in the narrow sense with one dimensional central distribution. The latter is integrable and its integral curves form a continuous foliation $W^c$ of $M$ with compact leaves (they are diffeomorphic to the unit circle; this foliation can be shown to be, indeed, Hölder continuous; see [42, 43]). There is a perturbation $G$ of $F$ which preserves volume and has nonzero Lyapunov exponents in the central direction (see [18]). In this case the central foliation is not absolutely continuous: for almost every $x \in M$ the conditional measure (generated by the Riemannian volume) on the leaf $W^c(x)$ of the central foliation passing through $x$ is atomic.

We describe a far more general version of this result. Let $(X, \nu)$ be a probability space and $f\colon X \to X$ an invertible transformation which preserves the measure $\nu$ and is ergodic with respect to $\nu$. Let also $M$ be a smooth compact Riemannian manifold and $\varphi\colon X \to \mathrm{Diff}^{1+\alpha}(M)$ be a map. Consider the skew-product transformation $F\colon X \times M \to X \times M$ given by

$$F(x, y) = (f(x), \varphi_x(y))$$

and assume that it is Borel measurable. We also assume that $F$ possesses an invariant ergodic measure $\mu$ on $X \times M$ such that $\pi_*\mu = \nu$ where $\pi\colon X \times M \to X$ is the projection.

Fix $x \in X$ and define the maps $\varphi_x^{(k)}$, $k \in \mathbb{Z}$, on $M$ by

$$\varphi_x^{(k+1)} = \varphi_{f^k(x)} \circ \varphi_x^{(k)},$$

where $\varphi_x^{(0)}$ is the identity map. Since the tangent bundle to $M$ is measurably trivial the derivative map of $\varphi$ along the $M$ direction gives a cocycle

$$\mathcal{A}\colon X \times M \times \mathbb{Z} \to GL(n, \mathbb{R}), \quad n = \dim M,$$

where $\mathcal{A}(x, y, k) = d_y \varphi_x^{(k)}$. If $\log^+ \|d\varphi\| \in L^1(X \times M, \mu)$ then the Multiplicative Ergodic Theorem and ergodicity of $\mu$ imply that the Lyapunov exponents of this cocycle are constant for $\mu$-almost every $(x, y)$,

$$\chi_1 < \cdots < \chi_\ell, \quad 1 \le \ell \le n.$$

**Theorem 10.3** (Ruelle and Wilkinson, [44]). *Assume that for some $\gamma > 0$ the function $\varphi$ satisfies*

$$\log^+ \|d\varphi\|_\gamma \in L^1(X, \nu), \tag{10.1}$$

*where $\|\cdot\|_\gamma$ is the $\gamma$-Hölder norm. Assume also that $\chi_\ell < 0$. Then there exists a set $S \subset X \times M$ of full measure and an integer $k \ge 1$ such that $\#(S \cap \{x\} \times M) = k$ for every $(x, y) \in S$.*

*Proof.* We need the following description of regular neighborhoods at typical points in $X \times M$.

**Lemma 10.4.** *There exists a set $\Lambda_0 \subset X \times M$ of full measure such that for any $\varepsilon > 0$ the following statements hold:*

(1) *there exist a measurable function $r\colon \Lambda_0 \to (0, 1]$ and a collection of embeddings $\Psi_{x,y}\colon B(0, r(x, y)) \to M$ such that $\Psi_{x,y}(0) = y$ and*

$$\exp(-\varepsilon) < \frac{r(F(x, y))}{r(x, y)} < \exp(\varepsilon);$$

(2) *if $\varphi_{(x,y)} = \Psi^{-1}_{F(x,y)} \circ \varphi_x \circ \Psi_{(x,y)}\colon B(0, r(x, y)) \to \mathbb{R}^n$ then*

$$\exp(\chi_1 - \varepsilon) \le \|d_0 \varphi^{-1}_{(x,y)}\|^{-1}, \ \|d_0 \varphi_{(x,y)}\| \le \exp(\chi_\ell + \varepsilon);$$

(3) *the $C^1$ distance $d_{C^1}(\varphi_{(x,y)}, d_0\varphi_{(x,y)}) < \varepsilon$ in $B(0, r(x, y))$;*

(4) *there exist $K > 0$ and a measurable function $A\colon \Lambda_0 \to \mathbb{R}$ such that for any $z, w \in B(0, r(x, y))$,*

$$\begin{aligned} K^{-1} \rho(\Psi_{(x,y)}(z), \Psi_{(x,y)}(w)) &\le \|z - w\| \\ &\le A(x) \rho(\Psi_{(x,y)}(z), \Psi_{(x,y)}(w)) \end{aligned}$$

*with*

$$e^{-\varepsilon} < \frac{A(F(x, y))}{A(x, y)} < e^{\varepsilon}.$$

Decomposing $\mu$ into a system of conditional measures

$$d\mu(x,y) = d\mu_x(y)d\nu(x)$$

and using invariance of $\mu$ with respect to $F$ we obtain for $\nu$-almost every $x \in X$,

$$(\varphi_x)_*\mu_x = \mu_{f(x)}.$$

The following result is an immediate consequence of Lemma 10.4.

**Lemma 10.5.** *There exists a set $\Lambda \subset \Lambda_0$ and real numbers $R > 0$, $C > 0$, and $0 < \alpha < 1$ such that*

(1) $\mu(\Lambda) > \frac{1}{2}$ *and if* $(x,y) \in \Lambda$ *then* $\mu_x(\Lambda_x) > \frac{1}{2}$ *where* $\Lambda_x = \{y \in M : (x,y) \in \Lambda\}$;
(2) *if* $(x,y) \in \Lambda$ *and* $z \in M$ *are such that* $\rho(y,z) \le R$, *then for all* $m \ge 0$,

$$\rho(\varphi_x^{(m)}(y), \varphi_x^{(m)}(z)) \le C\alpha^m \rho(y,z).$$

To prove the theorem it suffices to show that there is a set $B \subset X$ of positive $\nu$-measure such that for any $x \in B$ the measure $\mu_x$ has an atom. Indeed, for $x \in X$ set $d(x) = \sup_{y\in M} \mu_x(y)$. Clearly, the function $d(x)$ is measurable, $f$-invariant and positive on $B$. Since $f$ is ergodic we have $d(x) = d > 0$ for almost every $x \in X$. Let $S = \{(x,y) \in X \times M : \mu_x(y) \ge d\}$. Note that $S$ is $F$-invariant, has measure at least $d$ and hence, has measure 1. The desired result follows.

Let $\Lambda$ be the set constructed in Lemma 10.5 and $B = \pi(\Lambda)$. Clearly, $\nu(B) > 0$. We shall show that for any $x \in B$ the measure $\mu_x$ has an atom.

Let $\mathfrak{U}$ be a cover of $M$ by $N$ closed balls of radius $R/10$. For $x \in X$ set

$$m(x) = \inf \sum \operatorname{diam} U_j,$$

where the infimum is taken over all collections of closed balls $U_1, \dots, U_k$ in $M$ such that $k \le N$ and $\mu_x(\bigcup_{j=1}^k U_j) \ge \frac{1}{2}$. We also set $m = \operatorname{esssup}_{x\in B} m(x)$. We will show that $m = 0$. Otherwise, there is $p > 0$ such that

$$C\Delta N\alpha^p < \frac{m}{2}, \tag{10.2}$$

where $\Delta$ is the diameter of $M$. For $x \in B$ let $U_1(x)$, ..., $U_{k(x)}(x)$ be those balls in the cover $\mathfrak{U}$ that meet $\Lambda_x$. Since these balls cover $\Lambda_x$ and $\mu_x(\Lambda_x) > \frac{1}{2}$ we have that

$$\mu_x\left(\bigcup_{j=1}^{k(x)} U_j(x)\right) \ge \frac{1}{2}. \tag{10.3}$$

Since $(\varphi_x^{(i)})_* \mu_x = \mu_{f^i(x)}$ we obtain for all $i$ that

$$\mu_{f^i(x)}\left(\bigcup_{j=1}^{k(x)} \varphi_x^{(i)} U_j(x)\right) \geq \frac{1}{2}.$$

Since the balls $U_j(x)$ meet $\Lambda_x$ and have diameter less than $R/10$, we obtain, by Lemma 10.5, that

$$\operatorname{diam}(\varphi_x^{(i)} U_j(x)) \leq C\Delta\alpha^i. \tag{10.4}$$

Let $\tau(x)$ be the first return time of the point $x \in B$ to $B$ under the map $f^p$ and $B_i = \{x \in B : \tau(x) = i\}$. We have that $B = \bigcup_{i=1}^{\infty} B_i$ and since $f$ is invertible and its inverse $f^{-1}$ preserves $\nu$ we also have that

$$B' = \bigcup_{i=1}^{\infty} f^{pi}(B_i) = B \pmod 0.$$

If $z \in B'$ then $z = f^{pi}(x)$ for some $x \in B_i$ and some $i \geq 1$. It follows from the definition of $m(z)$ and inequalities (10.2), (10.3), and (10.4) that

$$m(z) \leq \sum_{j=1}^{k(x)} \operatorname{diam}(\varphi_x^{(pi)} U_j(x)) \leq Ck(x)\Delta\alpha^{pi} \leq CN\Delta\alpha^p \leq \frac{m}{2}.$$

This implies that

$$m = \operatorname{esssup}_{x\in B} m(x) = \operatorname{esssup}_{z\in B'} m(z) \leq \frac{m}{2}$$

contradicting the assumption that $m > 0$.

We conclude that $m = 0$ and hence, $m(x) = 0$ for $\nu$-almost every $x \in B$. This implies that for such an $x$ there is a sequence of closed balls $U^1(x)$, $U^2(x), \ldots$ with

$$\lim_{i\to\infty} \operatorname{diam} U^i(x) = 0$$

and $\mu_x(U^i(x)) \geq \frac{1}{2N}$ for all $i$. Take $z_i \in U^i(x)$. Any accumulation point of the sequence $\{z_i\}$ is an atom for $\mu_x$. $\square$

# References


[1] D. Anosov, *Geodesic flows on closed Riemannian manifolds with negative curvature*, Proc. Steklov Inst. Math. ,**90** (1969), 1–235.

[2] D. Anosov, Y. Sinai, *Certain smooth ergodic systems*, Russian Math. Surveys, **22** (1967), 103–167.

[3] L. Barreira, Ya. Pesin *Lyapunov Exponents and Smooth Ergodic Theory*, University Lecture Series, 23, AMS, Providence, RI, 2001.

[4] C. Bonatti, C. Matheus, M. Viana, A. Wilkinson, *Abundance of stable ergodicity*, Preprint (2003).

[5] M. Brin, *Hölder continuity of invariant distributions*, in *Smooth ergodic theory and its applications*, A. Katok, R. de la Llave, Ya. Pesin and H. Weiss eds., Proc. Symp. Pure Math., Amer. Math. Soc., 2001.

[6] M. Brin, *Topological transitivity of one class of dynamical systems and flows of frames on manifolds of negative curvature*, Func. Anal. Appl., **9** (1975), 9–19.

[7] M. Brin, *The topology of group extensions of C systems*, Mat Zametki, **18** (1975), 453–465.

[8] M. Brin, *On the dynamical coherence*, Ergod. Theory and Dyn. Syst., (2004).

[9] M. Brin, , PhD theses, Moscow, 1975.

[10] M. Brin, D. Burago, S. Ivanov *Partial hyperbolicity in dimension 3*, Preprint (2003).

[11] M. Brin, M. Gromov, *On the ergodicity of frame flows*, Invent. Math., **60** (1980), 1–7.

[12] M. Brin, H. Karcher, *Frame flows on manifolds of negative curvature*, Compos. Math., **52** (1984), 275–297.

[13] M. Brin, Ya. Pesin, *Partially hyperbolic dynamical systems*, Math. USSR-Izv., **8** (1974), 177–218.

[14] K. Burns, D. Dolgopyat, Ya. Pesin, *Partial hyperbolicity, Lyapunov exponents and stable ergodicity*, J. Stat. Phys., **109** (2002), 927–942.

[15] K. Burns, M. Pollicott, *Stable ergodicity and frame flows*, Preprint (2000).

[16] K. Burns, C. Pugh, A. Wilkinson, *Stable ergodicity and Anosov flows*, Topology, **39** (2000), 149–159.

[17] K. Burns, A. Wilkinson, *Stable ergodicity of skew products*, Ann. Sci. École Norm. Sup., **32** (1999), 859–889.

[18] D. Dolgopyat, *On dynamics of partially hyperbolic systems on three manifolds*, Preprint (1999).

[19] D. Dolgopyat, Ya. Pesin *Every Compact Manifold Carries a Completely Hyperbolic Diffeomorphism*, Ergod. Theory and Dyn. Syst., **22** (2002), 1–27.

[20] D. Dolgopyat, A. Wilkinson, *Stable accessibility is $C^1$ dense*, Asterisque (to appear)

[21] M. Grayson, C. Pugh, M. Shub, *Stably ergodic diffeomorphisms*, Ann. Math., **140** (1994), 295–329.

[22] N. Fenichel, *Persistence and smoothness of invariant manifolds of slows*, Indiana Univ. Math. J., **21** (1971–72), 193–226.

[23] J. Hadamard, *Sur l'itération et les solutions asymptotiques des équations différentielles*, Bull. Soc. Math. France **29** (1901), 224–228.

[24] B. Hasselblatt, *Regularity of the Anosov splitting and of Horospheric foliations* Ergod. Theory and Dyn. Syst., **14** (1994), no. 46, 45–666

[25] M. Hirsch, C. Pugh, M. Shub, *Invariant manifolds*, Bull. Amer. Math. Soc., **76** (1970), 1015–1019.

[26] M. Hirsch, C. Pugh, M. Shub, *Invariant manifolds*, Lecture Notes in Math., **583**, Springer-Verlag, 1977.

[27] M. Irwin, *On the stable manifold theorem*, Bull. London Math. Soc., **2** (1970), 196–198.

[28] M. Jiang, R. de Llave, Ya. Pesin, *On the integrability of intermediate distributions for Anosov diffeomorphisms*, Ergod. Theory and Dyn. Syst., **15** (1995), no. 2, 317–331.

[29] A. Katok, B. Hasselblatt, *Introduction to the modern theory of dynamical systems*, Cambridge University Press, Cambridge, 1995.

[30] A. Katok, A. Kononenko, *Cocycle stability for partially hyperbolic systems*, Math. Res. Letters, **3** (1996), 191–210.

[31] J. Mather, *Characterization of Anosov diffeomorphisms*, Nedel. Akad. Wetensch. Proc. Ser. A71, Indag. Math., **30** (1968), 479–483.

[32] J. Milnor, *Fubini foiled: Katok's paradoxical example in measure theory*, Math. Intelligencer **19** (1997), 30–32.

[33] O. Nikodym, *Sur la mesure des ensembles plans dont tous les points sont rectilinéairment accessibles*, Fund. Math., **10** (1927) 116–168.

[34] V. Niticǎ, A. Török, *An open dense set of stably ergodic diffeomorphisms in a neighborhood of a non-ergodic one*, Topology, **40** (2001), 259–278.

[35] D. Orendovici, Ya. Pesin *Chaos in Traveling Waves of Lattice Systems of Unbounded Media*, Proceedings of the IMA Volumes in Mathematics and its Applications, Numerical Methods for Bifurcation Problems and Large-scale Dynamical Systems, **119**, Springer-Verlag (1999), 327–359.

[36] O. Perron, *Über stabilität und asymptotishes Verhalten der Lösungen eines Systemes endlicher Differenzengleichungen*, J. Reine Angew. Math. **161** (1929), 41–64.

[37] Ya. Pesin, *On the existence of invariant fiberings for a diffeomorphism of a smooth manifold*, Matem. Sbornik, **91** (1973), no. 2, 202–210.

[38] J. Plante, *Anosov flows*, Amer. J. Math., **94** (1972), 729–754.

[39] C. Pugh, M. Shub, *Ergodicity of Anosov actions*, Invent. Math., **15** (1972), 1–23.

[40] C. Pugh, M. Shub, *Ergodic attractors*, Trans. Amer. Math. Soc., **312** (1989), no. 1, 1–54.

[41] C. Pugh (1-CA), M. Shub (1-IBM), *Stably ergodic dynamical systems and partial hyperbolicity. (English. English summary)*, J. Complexity, **13** (1997), no. 1, 125–179.

[42] C. Pugh, M. Shub, A. Wilkinson *Hölder foliations*, Duke Math. J. **86** (1997), no. 3 (1989), 517–546.

[43] C. Pugh, M. Shub, A. Wilkinson *Correction to: "Hölder foliations"*, Duke Math. J. **105** (2000), 105-106.

[44] D. Ruelle, A. Wilkinson, *Absolutely singular dynamical foliations*, Comm. Math. Phys., **219** (2001), 481–487.

[45] R. Sacker, *A perturbation theorem for invariant Riemannian manifolds*, Differential Equations and Dynamical Systems, Proc. Int. Symp., Mayaguez, P. R., (1967), 43–54.

[46] M. Shub, A. Wilkinson, *A stably Bernoullian diffeomorphism that is not Anosov*, Preprint (1998).

[47] M. Shub, A. Wilkinson, *Pathological foliations and removable zero exponents*, Invent. Math. (2001)

[48] S. Smale, *Differentiable dynamical systems*, Bull. Amer. Math. Soc., **73** (1967), 747–817.

[49] A. Talitskaya, *Stable ergodicity of suspension flows*, Preprint, PSU, 2004.

[50] H. Yamabe, *On an arcwise connected subgroup of a Lie group*, Osaka Math. J. **2** (1950), 13–14.

# Index